Quantitative Corpus Linguistics with R

"Today's linguists are increasingly called upon to master technical and statistical domains for which they do not usually receive primary training in their graduate programs. This clear and highly approachable introduction is an invaluable guide both for seasoned linguists moving in a more quantitative, corpus-based direction as well as for a new generation of graduate students just beginning work in this field. Gries' book strikes a useful balance between the theoretical and the technical aspects of corpus linguistics, while making the case for why linguists should devote more attention to corpora and statistical methods in their own research and why R is the ideal tool for the job."

– Steven Clancy, University of Chicago, USA

"This is the book I've always wished existed. It is the ultimate guide to the wonderful world of R."

– Dagmar Divjak, University of Sheffield, UK

"*Quantitative Corpus Linguistics with R* is a most welcome contribution to the rapidly expanding field of empirically-oriented research in linguistics. Written in a clear and highly accessible style by a respected authority in the field, the book introduces innovative research practices in a step-wise, interactive fashion. It will be of great interest and enormous benefit to both researchers and advanced students who are looking for ways to perfect the art of data retrieval and evaluation."

– Doris Schönefeld, Ruhr University Bochum, Germany

"*Quantitative Corpus Linguistics with R* is radically different from other corpus linguistics text-books. Gries' book offers a hands-on introduction to the methodological skills that are required to conduct sophisticated corpus linguistic analyses. An extremely useful book, this is the best practical introduction to the analysis of quantitative corpus data."

– Holger Diessel, Friedrich Schiller University of Jena, Germany

The first textbook of its kind, *Quantitative Corpus Linguistics with R* demonstrates how to use the open source programming language R for corpus linguistic analyses. Computational and corpus linguists doing corpus work will find that R provides an enormous range of functions that currently require several programs to achieve—searching and processing corpora, arranging and outputting the results of corpus searches, statistical evaluation, and graphing.

Stefan Th. Gries is Associate Professor of Linguistics at the University of California, Santa Barbara, USA.

Quantitative Corpus Linguistics with R
A Practical Introduction

Stefan Th. Gries

Routledge
Taylor & Francis Group

NEW YORK AND LONDON

First published 2009
by Routledge
270 Madison Ave, New York, NY 10016

Simultaneously published in the UK
by Routledge
2 Park Square, Milton Park, Abingdon, Oxon OX14 4RN

Routledge is an imprint of the Taylor & Francis Group, an informa business

© 2009 Taylor & Francis

Typeset in Minion by
RefineCatch Limited, Bungay, Suffolk
Printed and bound in the United States of America on acid-free paper by
Edwards Brothers, Inc.

Library of Congress Cataloging in Publication Data
Gries, Stefan Thomas, 1970–
    Quantitative corpus linguistics with R: a practical introduction / Stefan Th. Gries.
    p. cm.
    Includes bibliographical references and index.
    [etc.]
    1. Corpora (Linguistics)   2. Computational linguistics.   3. Linguistics—Statistical methods.   I. Title.
    P128.C68G75 2008
    410.1—dc22
    2008032294

ISBN10: 0–415–96271–4 (hbk)
ISBN10: 0–415–96270–6 (pbk)
ISBN10: 0–203–88092–7 (ebk)

ISBN13: 978–0–415–96271–1 (hbk)
ISBN13: 978–0–415–96270–4 (pbk)
ISBN13: 978–0–203–88092–0 (ebk)

Contents

Acknowledgments

I wish to thank the following people for input and support: Steven J. Clancy, Dagmar S. Divjak, Laura Kassner, and Gwillim Law for going over large parts of the book and useful and much appreciated suggestions for improvement, and Caroline V. David for discussion of an early draft of this book. Also, I am particularly grateful to two of the reviewers, one incredibly conscientious graduate student reviewer and one who turned out to be Randi Reppen, as well as to many students and participants of classes, summer schools, and workshops where parts of this book were used who provided much feedback over the last few years. Also, I thank Grabor Grothendieck for his incredibly useful R package gsubfn and much valuable input and discussion as well as Dani Bolliger and Heike Wagner for input regarding regular expressions and HTML. Last but certainly not least, I thank Stefanie Wulff for much feedback and invaluable support.

1
Introduction

1.1 Why Another Introduction to Corpus Linguistics?

This book is an introduction to corpus linguistics. If you are a little familiar with the field, this probably immediately triggers the question "Why yet another introduction to corpus linguistics?" This is a valid question because given the upsurge of corpus-linguistic studies there are also already quite a few introductions available. Do we really need another one? Predictably, I think the answer is "Yes", and the reason is that this introduction is radically different from every other introduction to corpus linguistics that's out there. For example, there are a lot of things that are regularly dealt with at length in introductions to corpus linguistics that I will only be concerned with very little:

- the history of corpus linguistics: Kaeding, Fries, early 1m word corpora, up to the contemporary mega corpora and the lively web-as-corpus discussion;
- how to compile corpora: size, sampling, balancedness, representativity, . . .;
- how to create corpus markup and annotation: lemmatization, tagging, parsing, . . .;
- kinds and examples of corpora: synchronic vs. diachronic, annotated vs. unannotated;
- what kinds of corpus-linguistic research other people have done.

One important characteristic of this book is that I would like to teach you *how* to do corpus linguistics. This is important since, to me, corpus linguistics is a method of analysis and thus talking about *how* to do things should enjoy a high priority.[1] Therefore, as opposed to reporting many previous studies, I will be more concerned with:

- aspects of data retrieval: how to generate frequency lists, concordances, collocation displays, etc. (don't worry if you do not know these terms, they will be explained later);
- aspects of data evaluation: how to save your results; how to import them into a spreadsheet program for further annotation; how to analyze results statistically; how to represent the results graphically; how to report your results.

A second important characteristic of this book is that it only uses open source software:

- R, the corpus linguist's all-purpose tool (cf. R Development Core Team 2008): a software that is a calculator, a statistics program, a (statistical) graphics program, and a programming language at the same time; the version used in this book is R 2.8.0 for Microsoft Windows and Ubuntu Intrepid Ibex, but all versions after 2.2 should work fine nearly all of the time;
- Tinn-R for Windows 2.1.1.3, a text editor that is particularly suited for writing scripts for, and interacting with, R;
- OpenOffice.org Calc 3.0.

The choice of these software tools, especially the decision to use R, has a variety of important implications, which should be mentioned early. As I just mentioned, R is a full-fledged programming language and, thus, a very powerful tool. However, this degree of power does come at a cost: in the beginning, it is undoubtedly more difficult to do things with R than with ready-made free or commercial concordancing software that has been written specifically for corpus-linguistic applications. For example, if you want to generate a frequency list or concordance of a word in a corpus with R, you must write a small *script* or a little bit of *code* in R's programming language, which is the technical way of saying, you write lines of text that are instructions to R. If you do not need pretty output, this script may actually consist of just two lines, but it could be more. On the other hand, if you have a ready-made concordancer, you click a few buttons (and enter a search term) to get the job done. You may therefore think, "Why go through the trouble of learning a programming language such as R?" It turns out, there is a variety of very good reasons for this, some of them are related to corpus linguistics, some are not.

First, the effort that goes into writing a script usually needs to be undertaken only once. As you will see below, once you have written your first few scripts while going through this book, you can usually reuse them for many different tasks and corpora, and the amount of time that is required to perform a particular task becomes identical to that of using a ready-made program. I freely admit, for example, that nearly all corpus-linguistic retrieval tasks in my own research are done with (somewhat adjusted) scripts or small snippets of code from this book. As a matter of fact, once you explore how to write your own functions, you can easily write your own versatile or specialized concordance functions yourself (I will make one such function available below). This way, the actual effort of doing a frequency list, concordance, or a collocate display often reduces to about the time you need with a concordance program. In fact, R may even be faster than competing applications: for example, some concordance programs read in the corpus files once before they are processed and then again for performing the actual task—R requires only one pass and may, therefore, outperform its competitors in terms of processing time.

Second, by learning to do your analyses with a programming language, you usually have more control over what you are actually doing: different concordance programs have different settings or different ways of handling searches that are not always obvious to the (inexperienced) user. As you will see in one of the assignments below, when you do a concordance of the string "perl" in the file <C:/_qclwr/_inputfiles/corp_perl.txt> with the default settings in the programs AntConc 3.2.1w, WordSmith Tools 4.0, and MonoConc Pro 2.2, then AntConc finds 253 matches whereas WordSmith Tools and MonoConc Pro 2.2 find 248 matches. Users then not only face the problem of what to do with these conflicting results, but are then basically required to figure out why the counts differ or, put differently, how these programs have defined what a word is and how you can change their settings, etc. With a programming language, *you* are in the driver's seat: you define what a word is, and you define how a particular search is implemented for a particular search word and a particular corpus, and your results will be maximally replicable.

Third, if you use a particular concordancing software, you are at the mercy of its developer. If the developers change its behavior, its results output, or its default settings, you can only hope that this does not affect your results. Worse even, developers might even discontinue the development altogether . . . But there are also additional important advantages of the fact that R is an open source tool/programming language. For instance, there is a large number of functions that are contributed by users all over the world and often allow effective shortcuts that are not or hardly possible with ready-made applications, which you cannot tweak as you wish, and contrary to commercial concordance software, bug-fixes are available very fast.

Fourth and related to that, a programming language is a much more versatile tool than any ready-made software application. You will see many examples below where R provides more precise and/or customizable output than most available concordancers, or where R allows you to do things that regular concordance software could only do with a lot of manual effort or, more commonly, not at all. For example, there are many corpora or databases that come in formats that ready-made concordancers cannot handle at all but which, with just a small script, are easy to deal with in R. One case in point is the CELEX database, which provides a wealth of phonological information on English, German, and Dutch words, but comes in a format that is difficult or even impossible to use with spreadsheet software or concordancing programs but which is easy to handle in R (cf. Gries 2004b, 2006b for examples); another one is that many concordancers, even commercial ones, can't handle Unicode yet and are thus severely limited in terms of the languages they can be used with— R has no such problems (at least on MacOS X and Linux; Windows is unfortunately a bit of a different story). In a way, once you have mastered the basic mechanisms, there is basically no limit to what you can do with it both in terms of linguistic processing and statistical evaluation.

Fifth and also related to that, the knowledge you will acquire here is less specialized: you will not be restricted to just one particular software application (or even one version of one particular software application) and its restricted set of features. Rather, you will acquire knowledge of a programming language and regular expressions that will allow you to use many different utilities and to understand scripts in other programming languages, such as Perl or Python. (At the same time, R is simpler than Perl or Python, but can also interface with them via RSPerl and RSPython respectively; cf. <http://www.omegahat.org/>.) For example, if you ever come across scripts by other people or decide to turn to these languages yourself, you will benefit from knowing R in a way that no ready-made concordancing software would allow for. If you are already a bit familiar with corpus-linguistic work, you may now think, "But why turn to R and not use Perl or Python (especially since you say Perl and Python are similar anyway and many people already use one of these languages)?" This is a good question, and I can even tell you that sometimes Perl is even faster than R and I myself used Perl for a few applications (cf. Danielsson 2004 for a brief introduction to corpus linguistics using Perl). However, I think I also have a good answer why to use R instead. First, the issue of speed is much less of a problem than one may think. R is fast enough for most applications anyway and usually so stable that you can simply execute a script while you are in class or even over the night and collect the results afterwards. Second, R has other advantages. One is that, in addition to the text processing capabilities, R also offers a large number of ready-made functions for the statistical evaluation and graphical representation of data, which allow you to perform just about *all* corpus-linguistic tasks within only one programming environment: you can do your data processing, data retrieval, annotation, statistical evaluation, graphical representation . . . *everything* within just one environment whereas if you wanted to do all these things in Perl or Python, you would require a huge amount of separate programming. Consider a very simple example: R has a function called `table` that generates a frequency table. To perform the same in Perl you would either have to have a small loop counting elements in an array and in a stepwise fashion increment their frequencies in a hash or, more cleverly, to program a subroutine that you would then always call upon. While this is no problem with a one-dimensional frequency list, this is much harder with multidimensional frequency tables: Perl's arrays of arrays or hashes of arrays, etc. are not for the faint-hearted while R's `table` is easy to handle and additional functions (e.g., `xtabs`, `ftable`, . . .) allow you to handle such tables very easily. I believe learning one environment is sufficiently hard for beginners and therefore recommend using the more comprehensive environment with the greater number of simpler functions. And, once you have mastered the fundamentals of R and face situations where you do need maximal computational power, switching to Perl in a limited number of cases will be easier for you anyway, especially since much of the programming language's syntax is similar and the regular expressions used in this book are all Perl compatible. (Let me tell you, though, that in all my years using R, there were a mere two instances where I had to switch to Perl.)

A final, very down-to-earth advantage of using open source software is of course that it comes free of charge. Any student or any department's computer lab can afford it without expensive licenses, temporally limited or functionally restricted licenses, or irritating nag screens. All this, hopefully, makes a strong case for the choice of software.

1.2 Outline of the Book

This book is structured as follows. The next chapter, Chapter 2, defines the notion of a corpus and provides a brief overview of what I consider to be the three most central corpus-linguistic methods, namely frequency lists, collocations, and concordances. Chapter 3 then introduces you to the fundamentals of R, covering a variety of functions from different domains, but the area which receives most consideration is that of text processing. Chapter 4 illustrates how the methods introduced in Chapter 3 are applied to corpus data in the ways outlined in Chapter 2. Using a variety of different kinds of corpora, you will learn in detail how to write your own small programs in R for corpus-linguistic analysis. Chapter 5 introduces you to some fundamental aspects of statistical thinking and testing. The questions to be covered in this chapter are, What are hypotheses? How do I check whether my results are noteworthy?, etc.

Chapter 6 introduces you to a variety of applications. When you download all necessary files from the book's companion website, you will also download a set of case studies, each with between one and six assignments. These case studies introduce you to more realistic and, sometimes, more demanding corpus-linguistic applications. Some of these are based on published research; others exhibit features that are useful for corpus-linguistic studies and that are sometimes offered by other (commercial) software programs. The focus, however, is not on these other studies or programs—these studies serve as templates on the basis of which you get assignments and problems to solve by writing your own scripts to replicate the findings reported in a study or prepare the data for an independent project. As such, these case studies are from a wide array of applications: there are examples from morphology, syntax, semantics, and text linguistics; apart from concordancing, etc., you learn how to identify statistically significant key words in texts, use R to automatically access websites as corpus data, compute mean lengths of utterances for language acquisition research, and much more. In addition to the actual case studies, I also point you to a variety of domains of research, to which you can also apply the skills you learn from this book. Last, the appendices contain links to relevant websites, answer keys to all exercise boxes, and solutions to all assignments.

Before we begin, a few short comments on the nature of this book are necessary. As I already mentioned, this introduction to corpus linguistics is different from every other introduction I know—actually, it is the kind of introduction I always wanted to buy and could never find. This has two consequences. On the one hand, this book is not a book that requires much previous knowledge: it neither presupposes linguistic knowledge other than what you learn in your first introduction to linguistics nor mathematical or any programming knowledge.

On the other hand, this book is an attempt to teach you a lot about how to be a good corpus linguist. As a good corpus linguist, you have to combine many different concrete methodological skills (*and* many equally important analytical skills that I will not be concerned with here). Many of these methodological skills are addressed here, such as some very basic knowledge of computers (operating systems, file types, etc.), data management, regular expressions, some elementary programming skills, some elementary knowledge of statistics, etc. What you must know therefore is that (i) nobody has ever learned all this just by reading—you must *do* things—and (ii) this is *not* an easy book that you can read 10 minutes at a time in bed before you fall asleep. What these two things mean is that you should read this book while sitting at your computer and directly entering the code and working on the examples. You will need practice to master all the concepts introduced here, but will be rewarded by acquiring skills that give you access to a variety of data and approaches you may not have considered accessible to you—at least that's what happened to me. Undergraduates in my corpus classes without programming experience have quickly learned to write small programs that do things better than many concordancing programs, and you can do the same.

In order to facilitate your learning process, there are four different ways in which I try to help you get more out of this book. First, there are small think breaks. These are small assignments that you should try to complete before you read on, and answers to them follow immediately in the text. Second, there are exercise boxes with small assignments. Ideally, you should complete them and check your answers in the answer key before you read any further, but it is not always necessary to complete them to understand what follows so you can also return to them later at your own leisure. Third, there are many boxes with recommendations for further study/exploration. With perhaps one or two exceptions, they are just that, recommendations, which means you do not need to follow up on them to understand later material in the book—I just encourage you to follow them up anyway. Fourth and in addition to the above, I would like to mention that the companion website for this book is hosted at a Google group called "CorpLing with R", which I created and maintain. I would like to encourage you to go to the website of the newsgroup (the URL can be found in the Appendix) and become a member. Not only do you have to go to the companion website anyway to get all the files that belong to this book, but if you become a member of the list:

- you can send questions about corpus linguistics with R to the list and, hopefully, get useful responses from some kind soul(s);
- post suggestions for revisions of this book there;
- inform me and the other readers of errors you found and, of course, be informed when other people or I myself find errata.

Thus, while this is not an easy book, I hope these aids help you to become a good corpus linguist. If you work through the whole book, you will be able to do many things you could not even do with commercial concordancing software; many of the scripts you find here are taken from actual research, are in fact simplified versions of scripts I have used myself for published papers. In addition, if you also take up the many recommendations for further exploration that are scattered throughout the book you will probably find ever new and more efficient ways of application.

1.3 Recommendation for Instructors

If this book is not used for self-study but in a course, I would recommend using it in a two-quarter or two-semester sequence such that

- in the first quarter or semester, the course deals with the contents of Chapters 1 to 5, supplemented with additional readings if required;
- in the second quarter or semester, the students deal with assignments of Chapter 6, either all of them or, more likely, assignments that fit each student's personal interest and discuss them with the instructor.

Another possibility is to deal with Chapters 1 to 4 in the course of a quarter or semester and assign (parts of the) assignments in Chapter 6 as small homework assignments. Obviously, however, many more arrangements are conceivable.

2

The Three Central Corpus-linguistic Methods

This last point leads me, with some slight trepidation, to make a comment on our field in general, an informal observation based largely on a number of papers I have read as submissions in recent months. In particular, we seem to be witnessing as well a shift in the way some linguists find and utilize data—many papers now use corpora as their primary data, and many use internet data.

(Joseph 2004: 382)

In this chapter, you will learn what a corpus is (plural: corpora) and what the three methods are to which nearly all corpus-linguistic work can be reduced.

2.1 Corpora

Before we start to actually look at corpus linguistics, we have to clarify our terminology a little. In this book, I will distinguish between three different things: a corpus, a text archive, and an example collection.

2.1.1 What is a Corpus?

In this book, the notion of "corpus" refers to a machine-readable collection of (spoken or written) texts that were produced in a natural communicative setting, and the collection of texts is compiled with the intention (1) to be representative and balanced with respect to a particular linguistic variety or register or genre and (2) to be analyzed linguistically. The parts of this definition need some further clarification themselves:

- "machine-readable" refers to the fact that nowadays virtually all corpora are stored in the form of plain ASCII or Unicode text files that can be loaded, manipulated, and processed platform-independently. This does not mean, however, that corpus linguists only deal with raw text files—quite the contrary: some corpora are shipped with sophisticated retrieval software that makes it possible to look for precisely defined syntactic and/or lexical patterns. It does mean, however, that you would have a hard time finding corpora on paper, in the form of punch cards or digitally in HTML or Microsoft Word document formats; the current standard is text files with XML annotation.

- "produced in a natural communicative setting" means that the texts were spoken or written for some authentic communicative purpose, but not for the purpose of putting them into a corpus. For example, many corpora consist to a large degree of newspaper articles. These meet the criterion of having been produced in a natural setting because journalists write the article to be published in newspapers and to communicate something to their readers, but not because they want to fill a linguist's corpus. Similarly, if I obtained permission to record all of a particular person's conversations in one week, then hopefully, while the person and his interlocutors usually are aware of their conversation being recorded, I will obtain authentic conversations rather than conversations produced only for the sake of my corpus.

- I use "representative [. . .] with respect to a particular variety" here to refer to the fact that the different parts of which the linguistic variety I am interested in are all manifested in the corpus. For example, if I was interested in phonological reduction patterns of speech of adolescent Californians and recorded only parts of their conversations with several people from their peer group, my corpus would not be representative in the above sense because it would not reflect the fact that some sizable proportion of the speech of adolescent Californians may also consist of dialogs with a parent, a teacher, etc., which would therefore also have to be included.

- I use "balanced with respect to a particular linguistic variety" to mean that not only should all parts of which a variety consists be sampled into the corpus, but also that the proportion with which a particular part is represented in a corpus should reflect the proportion the part makes up in this variety and/or the importance of the part in this variety. For example, if I know that dialogs make up 65 percent of the speech of adolescent Californians, approximately 65 percent of my corpus should consist of dialog recordings. This example already shows that this criterion is more of a theoretical ideal: How would one measure the proportion that dialogs make up of the speech of adolescent Californians? We can only record a tiny sample of all adolescent Californians, and how would we measure the proportion of dialogs—in terms of time? in terms of sentences? in terms of words? And how would we measure the importance of a particular linguistic variety? The implicit assumption that conversational speech is somehow the primary object of interest in linguistics also prevails in corpus linguistics, which is why many corpora aim at including as much spoken language as possible, but on the other hand a single newspaper headline read by millions of people may have a much larger influence on every reader's linguistic system than twenty hours of dialog. In sum, balanced corpora are a theoretical ideal that corpus compilers constantly bear in mind, but the ultimate and exact way of compiling a balanced corpus has remained mysterious so far.

Many people consider a text archive to be something different, namely a database of texts

- that may not have been produced in a natural setting;
- that has often not been compiled for the purposes of linguistic analysis;
- that has often not been intended to be representative with respect to a particular linguistic variety or speech community;

- that has often not been intended to be balanced with respect to a particular linguistic variety or speech community.

However, the distinction between the corpora and text archives is often blurred. It is theoretically easy to make, but in practice often not adhered to very strictly. For example, if a publisher of a popular computing periodical makes all the issues of the previous year available on his website, then the first criterion is met, but not the last three. However, because of their availability and their size, many corpus linguists use them as resources, and as long as one bears their limitations in terms of representativity, etc. in mind, there is little reason not to do that.

Finally, an *example collection* is just what the name says it is, namely a collection of examples that, typically, the person who compiled the examples came across and noted down. For example, much psycholinguistic research in the 1970s was based on collections of speech errors compiled by the researchers themselves and/or their helpers. Occasionally, people refer to such collections as error corpora but we will not use the term *corpus* for these. It is easy to see how such collections compare to corpora. On the one hand, for example, an error collection may fail to be representative and balanced because some errors—while occurring frequently in authentic speech—are more difficult to perceive than others and thus hardly ever make it into a collection. This would be an analog to the balancedness problem outlined above. On the other hand, the perception of errors is contingent on the acuity of the researcher while, with corpus research, the corpus compilation would not be contingent on a particular person's perceptual skills. Finally, because of the scarcity of speech errors, usually all speech errors perceived (in a particular amount of time) are included into the corpus whereas, at least usually and ideally, corpus compilers are more picky and select the material to be included with an eye to the criteria of representativity and balancedness outlined above.[1] Be that as it may, for the sake of terminological clarity, we will carefully distinguish the notions of corpora, text archives, and example collections and focus only on the first category.

2.1.2 What Kinds of Corpora are There?

Corpora differ in a variety of ways. There are a few distinctions you should be familiar with if only to be able to find the right corpus for what you want to investigate. The most basic distinction is that between *general corpora* and *specific corpora*. The former intend to be representative and balanced for a language as a whole—within the above-mentioned limits, that is—while the latter are by design restricted to a particular variety, register, genre, . . .

Another important distinction is that between *raw corpora* and *annotated corpora*. Raw corpora consist of files only containing the corpus material (cf. (1)) while annotated corpora in addition also contain additional information. Annotated corpora are very often annotated according to the standards of the Text Encoding Initiative (TEI) or the Corpus Encoding Standard (CES), and then have two parts. The first part is called the header, which provides information that is typically characterized as markup. This is information about (i) the text itself, e.g. where the corpus data come from, which language is represented in the file? which (part of a) newspaper or book has been included? who recorded whom where and when? who has the copyright? which annotation comes with the file? and information about (ii) its formatting, printing, processing, etc. Markup refers to objectively codable information—the fact that there is a paragraph in a text or that a particular speaker is female can be made without doubt. This information helps users to quickly determine whether a particular file is part of the register one wishes to investigate or not.

The second part is called the body and contains the corpus data proper—i.e. what people actually said or wrote—as well as linguistic information that is usually based on some linguistic theory: parts of speech or syntactic patterns, for example, can be matters of debate. In what follows

I will briefly (and non-exhaustively!) discuss and exemplify a few common annotation schemes (cf. Leech's and McEnery and Xiao's articles in Wynne 2005 as well as McEnery, Xiao, and Tono 2006: A.3 and A.4 for more discussion).

First, a corpus may be *lemmatized* such that each word in the corpus is followed by its lemma, i.e. the form under which you would look it up in a dictionary (cf. (2)). A corpus may have so-called *part-of speech tags* so that each word in the corpus is followed by an abbreviation giving the word's part of speech (POS) and sometimes also some morphological information (cf. (3)). A corpus may also be phonologically annotated (cf. (4)). Then, a corpus may be *syntactically parsed*, i.e. each word is followed by an abbreviation giving the word's syntactic status (cf. (5)). Finally and as a last example, a corpus may contain several different annotations on different lines (or tiers) at the same time, a format especially common in language acquisition corpora (cf. (6)).

(1)   I did get a postcard from him.

(2)   I<I> did<do> get<get> a<a> postcard<postcard> from<from> him<he>.<punct>

(3)   I<PersPron> did<VerbPast> get<VerbInf> a<Det> postcard<NounSing> from<Prep> him<PersPron>.<punct>

(4)   [@:] • I • ^did • get • a • !p\ostcard • fr/om • him# • - • -

(5)   <Subject, • NP>
          I<PersPron>
      <Predicate, • VP>
          did<Verb>
          get<Verb>
          <DirObject, • NP>
              a<Det>
              postcard<NounSing>
          <Adverbial, • PP>
              from<Prep>
              him<PersPron>.

(6)   *CHI: I did get a postcard from him
      %mor:
      pro|I • v|do&PAST • v|get • det|a • n|postcard • prep|from • pro|him • .
      %lex: get
      %syn: trans

Other annotation includes annotation with regard to semantic characteristics, stylistic aspects, anaphoric relations (coreference annotation), etc. Nowadays, most corpora come in the form of SGML or XML files, and we will explore many examples involving these formats in the chapters to come. As is probably obvious from the above, annotation can sometimes be done completely automatically (possibly with human error checking), semi-automatically, or must be done completely manually. Part-of-speech tagging, probably the most frequent kind of annotation, is usually done automatically, and taggers are claimed to achieve accuracy rates of 97 percent, a number that I sometimes find hard to believe when I look at corpora, but that is a different story.

Then, there is a difference between *diachronic corpora* and *synchronic corpora*. The former aims at representing how a language/variety changes over time while the latter provides, so to speak, a snapshot of a language/variety at one particular point of time. Yet another distinction is that between *monolingual corpora* and *parallel corpora*. As you might already guess from the names, the former have been compiled to provide information about one particular language/variety, etc. whereas the latter ideally provide the same text in several different languages. Examples include translations from EU Parliament debates into the 23 languages of the European Union or the Canadian Hansard corpus, containing Canadian Parliament debates in English and French. Again ideally, a parallel corpus does not just have the translations in different languages, but has the translations sentence-aligned, such that for every sentence in language L_1, you can automatically retrieve its translation in the languages L_2 to L_n.

The next distinction to be mentioned here is that of *static corpora* vs. *dynamic/monitor corpora*. Static corpora have a fixed size (e.g. the Brown corpus, the LOB corpus, the British National Corpus) whereas dynamic corpora do not, since they may be constantly extended with new material (e.g. the Bank of English).

The final distinction I would like to mention at least briefly involves the encoding of the corpus files. Given especially the predominance of work on English in corpus linguistics, until rather recently, many corpora came in the so-called ASCII (American Standard Code for Information Interchange) character encoding, an encoding scheme that encodes $2^7 = 128$ characters as numbers and that is largely based on the English characters. With these characters, special characters that were not part of the ASCII character inventory (e.g. the "é" was paraphrased as "é"). However, the number of corpora for many more languages has been increasing steadily and given the large number of characters that writing systems such as Chinese have, this is not a practical approach, language-specific character encodings were developed (e.g. ISO 8859-1 for Western European Languages vs. ISO 2022 for Chinese/Japanese/Korean languages). However, in the interest of over-coming compatibility problems that arose because now different languages used different character encodings, the field of corpus linguistics is currently moving towards using only one unified (i.e. not language-specific) multilingual character encoding, Unicode (most notably UTF-8). This development is in tandem with the move towards XML corpus annotation and, more generally, UTF-8's becoming the most widely used character encoding on the WWW (cf. <http://google blog.blogspot.com/2008/05/moving-to-unicode-51.html>).

Now that you know a bit about the kinds of corpora that are out there, there is one other really important point to be made. While we will see below that corpus linguistics has a lot to offer to the analyst, it is worth pointing out that, strictly speaking at least, the only thing corpora can provide is information on frequencies. Put differently, there is no meaning in corpora, and there are no functions in corpora—corpora only contain

- frequencies of occurrence, i.e. how often morphemes, words, grammatical patterns, etc. occur in (parts of) a corpus; or
- frequencies of co-occurrence of the same kinds of items, i.e. how often do morphemes occur with particular words? how often do particular words occur in a certain grammatical construction? etc.

. . . and it is up to the researcher to interpret frequencies of occurrence and co-occurrence in meaningful or functional terms. The assumption underlying basically all corpus-based analyses, however, is that, since formal differences correspond to functional differences, different frequencies of (co-)occurrences of formal elements are supposed to reflect functional regularities. "Functional" is understood here in a very broad sense as anything—be it semantic, discourse-pragmatic, . . .—that is intended to perform a particular communicative function. On a very general level, the frequency information a corpus offers is exploited in three different ways, which will be the subject of this chapter: frequency lists (cf. Section 2.2), lexical co-occurrence lists or collocations (cf. Section 2.3), and concordances (cf. Section 2.4).

2.2 Frequency Lists

The most basic corpus-linguistic tool is the frequency list. You generate a frequency list when you want to know how often words occur in a corpus. Thus, a frequency list of a corpus is usually a two-column table with all words occurring in the corpus in one column and the frequency with which they occur in the corpus in the other column. Since the notion of *word* is a little ambiguous here, it is useful to introduce a common distinction between (word) type and (word) token. The string "the word and the phrase" contains five (word) tokens ("the", "word", "and", "the", and "phrase"), but only four (word) types ("the", "word", "and", and "phrase"), of which one occurs twice. In this parlance, a frequency list lists the types in one column and their token frequencies in the other. Typically, one out of three different sorting styles is used: frequency order (ascending or, more typically, descending; cf. the left panel of Table 2.1), alphabetical (ascending or descending), and occurrence (each word occurs in a position reflecting its first occurrence in the corpus).

Apart from this simple form, there are two other varieties of frequency lists that are sometimes found. On the one hand, a frequency list may provide the frequencies of all words together with the words with their letters reversed. This may not seem particularly useful at first, but even a brief look at the central panel of Table 2.1 clarifies that this kind of display can sometimes be very helpful because it groups together words that share a particular suffix, here the adverb marker -*ly*. On the other hand, a frequency list may not list each individual word token and their frequencies but so-called *n*-grams, i.e. sequences of *n* words such as bigrams (where *n* = 2, i.e. word pairs) or trigrams (where *n* = 3, i.e. word triples) and their frequencies; cf. the right panel of Table 2.1 for an example involving bigrams.

Table 2.1 Examples of differently ordered frequency lists

Words	Freq	Words	Freq	Bigrams	Freq
the	62,580	yllufdaerd	80	of the	4,892
of	35,958	yllufecaep	1	in the	3,006
and	27,789	yllufecarg	5	to the	1,751
to	25,600	yllufecruoser	8	on the	1,228
a	21,843	yllufeelg	1	and the	1,114
in	19,446	yllufeow	1	for the	906
that	10,296	ylluf	2	at the	832
is	9,938	yllufepoh	8	to be	799
was	9,740	ylluferac	87	with the	783
for	8,799	yllufesoprup	1	from the	720

Frequency lists are sometimes more problematic than they seem, because they presuppose that the linguist (and/or his computer program) has a definition of what a word is and that this definition is shared by other linguists (and their computer programs). This need not be the case, however, as the following exercise will demonstrate.

Exercise box 2.1: What is a word, or what does your computer think a word is?

(1)   Write up a plain English definition of how you would "tell a computer program" what a word is. When you are finished, do the next exercise.

(2)   How does your definition handle the expressions *better-suited* and *ill-defined*? How does it handle *armchair-linguist* and *armchair linguist* and *armchair-linguist's*? How does it handle *http://www.linguistics.ucsb.edu*? How does it handle *This* and *this*? How does it handle *1960* and *25-year-old*? And *favor* and *favour*? And *Peter's car*, *Peter's come home*, and *Peter's sick*? And what about позволено (Cyrillic) or 宜蘭童玩節停辦 (Chinese) in an English text?

Bear in mind, therefore, that you need to exercise caution in comparing frequency lists from different sources.[2] Another noteworthy aspect is that frequency lists are often compiled so as not to include words from a so-called stop list. For example, one can often exclude a variety of frequent function words such as *the, of, and,* etc., because these are often semantically not particularly revealing since they occur nearly indiscriminately throughout the corpus.[3]

For what follows below, it is useful to introduce the distinction between a lemma and a word-form: *go, goes, going, went,* and *gone* are all different word forms even though they belong to the same lemma, namely *go*; in the remainder of this chapter, I will only use *word* where the difference between *lemma* and *word form* is not relevant. Computer programs normally define word forms as a sequence of alphabetic (or alphanumeric) characters uninterrupted by whitespace (i.e. spaces, tabs, and newlines) and allow the user to specify what to do with hyphens, apostrophes, and other special characters.

Frequency lists are useful for a variety of purposes. For example, much contemporary work in usage-based linguistics assumes that the type and token frequencies of linguistic expressions are correlated with the degree of cognitive entrenchment of these expressions, and studies such as Bybee and Scheibman (1999) and Jurafsky et al. (2000) related frequencies of expressions to their readiness to undergo processes of phonological reduction and grammaticalization. For example, Bybee and Scheibman (1999) show that the vowel in *don't* is more likely to be reduced to schwa in frequent expressions such as *I don't know* (as opposed to, say, *They don't integrate easily*). In psycholinguistics, models of language production have long incorporated frequency effects, but there is now a growing recognition that information of percentages, conditional probabilities, etc. of all kinds are represented in the linguistic systems of speakers and, thus play a primary role at all levels of linguistic analysis. On the more practical side of things, word frequency lists are useful to choose experimental stimuli correctly: for example, the logs of frequency of words are related to psycholinguistic measures such as reaction times to these words (in priming or naming studies), which often makes it necessary to control for such effects in order not to bias your reaction times. Also, you may want to do a study on tip-of-the-tongue states and, thus, make sure the words you use are sufficiently infrequent.

A much more applied example which we will also be concerned with below involves the identification of words that are (significantly) overrepresented in one corpus as compared to a balanced reference corpus. This can be used to identify the words that are most characteristic of a particular domain: a word that is about equally frequent in a particular corpus and a larger reference corpus is probably not very revealing in terms of what the smaller corpus is about. However, if a corpus contains the words *economic, financial, shares, stocks* much more often than a balanced reference corpus, guessing the topics covered in the smaller corpus is straightforward.

This kind of approach can be used to generate lists of important vocabulary for language learners.

In the domain of natural language processing or computational linguistics, the frequency of items is relevant to, among other things, speech recognition. For example, imagine a computer gets ambiguous acoustic input in a noisy environment and tries to recognize which word is manifested in the input. If the computer cannot identify the input straightforwardly, one (simplistic) strategy would be for it to assume that the word it hears is the most frequent one of all those that are compatible with the acoustic input. Another area of interest is, for example, spelling error correction, where frequency lists can be useful in two ways: first, for the computer to recognize that a string is probably a typo because it does not occur in a gold standard frequency list of words of the input language; second for the computer to rank suggestions for correction such that the computer first determines a set of words that are sufficiently similar to the user's input and then ranks them according to their similarity to the input and their frequency. From a methodological perspective, frequency lists are useful for computing many co-occurrence statistics. Finally, as will be obvious from the set of reading recommendations below, frequency lists may reflect sociocultural differences.

2.3 Lexical Co-occurrence: Collocations

One of the most central methodological concepts is that of co-occurrence. Corpus linguists have basically been concerned with three different kinds of co-occurrence phenomena:

- collocation: the probabilistic co-occurrence of word forms such as *different from* vs. *different to* vs. *different than*; or the absolute frozenness of expressions such as *kith and kin* or *by and large*;
- colligation: the co-occurrence of word forms with grammatical phenomena such as part-of-speech categories, grammatical relations, or definiteness such as the preference of *consequence* to occur as a complement (but not an adverbial) and with an indefinite article;
- (grammar) patterns or collostructions: the co-occurrence of words/lemmas with morpho-syntactic patterns or constructions such as the ditransitive construction or the cleft construction such as the preference of *to hem* to occur in the passive; or the association of the ditransitive to forms of the verb *to give*.

In this section, we will restrict ourselves to collocations because they are a natural extension of frequency lists.

Collocations are co-occurrences of words, which are then referred to as *collocates*. Often, one uses the letters L and R (for left and right) together with a number indicating the position of one collocate with respect to the other to talk about collocations. For example, if you call up a concordance of the word *difference*, then you will most certainly find that the most frequent L1 collocate is *the* while the most frequent R1 collocate is *between*. Thus, a collocate display for a word tells you which other words occur at different positions around the target word and how frequently. A *collocate* display for a word, accordingly, is usually a table with several columns such that each column lists the collocates of the target word at a particular position together with their frequencies. In other words, a collocate display is a list of frequency lists for particular positions around a word. To look at a simple example with two adjectives, consider the two adjectives *alphabetic* and *alphabetical*.

Table 2.2 is a collocate display of *alphabetic*; the strings in angular brackets are word class tags (cf. the Appendix for pointers to lists of all tags used in this book), and Table 2.3 is a collocate

Table 2.2 A collocate display of *alphabetic* based on the BNC

Word at L1	Freq L1	Node word	Freq Node	Word at R1	Freq R1
<w prf>of	8	<w aj0>alphabetic	42	<w nn1>literacy	7
<w at0>the	6			<w nn1>writing	5
<w at0>an	5			<w nn1>order	3
<w prp>in	2			<w nn1>character	3
<w prp>such as	2			<w cjc>and	2
<w dps>our	2			<w nn1>system	2
<w cjs>when	2			<w nn2>characters	2
<w aj0 widespread>	1			<w nn1>culture	2
<w nn2>systems	1			<w prp>in	1
<w aj0>varying	1			<c pun>.	1

Table 2.3 A collocate display of *alphabetical* based on the BNC

Word at L1	Freq L1	Node word	Freq Node	Word at R1	Freq R1
<w prp>in	77	<w aj0>alphabetical	234	<w nn1>order	89
<w at0>an	36			<w nn1>index	15
<w at0>the	23			<w nn1>list	13
<w prf>of	6			<w nn1>indexing	12
<w cjc>and	6			<w nn1>subject	12
<c pun>.	6			<w nn1>sequence	11
<c pun>,	6			<w nn1>listing	9
<w aj0>ascending	5			<w nn1>guest	6
<w cjc>or	5			<w cjc>and	5
<w aj0>strict	4			<w nn1>description	2

display of *alphabetical*; both are based on data from the British National Corpus World Edition (BNC).

Note that a collocate display is read vertically: Row 1 of Table 2.2 does not tell you anything about the string *of alphabetic literacy*—it tells you that *of* is the most frequent word at L1 (occurring there eight times) and that *writing* is the second most frequent word at R1 (with five occurrences). Now, if we compare the R1 collocates of *alphabetic* and *alphabetical*, can you already discern a tendency of the specific semantic foci of the two adjectives?

THINK

BREAK

The difference between the two adjectives can probably be paraphrased easiest by stating what the opposites of the two adjectives are. My suggestion would be that the opposite of *alphabetic* is *numeric* whereas the opposite of *alphabetical* is *unordered*, but a more refined look at the data may reveal a more precise picture.

Collocate displays are an important tool within semantics and lexicography (cf. Sinclair 1987, Stubbs 1995, Kilgarriff and Tugwell 2001), as well as language teaching (cf. Lewis 2000); the phenomenon of near synonymy is a particularly fruitful area of application. Consider for example the English adjectives *fast*, *quick*, *rapid*, and *swift*. While we would be hard-pressed to intuitively identify the difference between these semantically extremely similar expressions, inspecting the position R1 in collocate displays for these adjectives would give us a very good clue as to the nouns these adjectives usually modify attributively and exhibit semantic or other distributional regularities.

Another area of application is what has been referred to as semantic prosody, i.e. the fact that collocates of some word *w* may imbue *w* with a particular semantic aura even though this aura is not part of the semantics of *w* proper. One of the standard textbook examples is the English verb *to cause*. As you probably sense intuitively, *to cause* primarily, though not exclusively, collocates with negative things (*problem, damage, harm, havoc, distress, inconvenience*, etc.) although causing as such need not be negative. This is not only a theoretically interesting datum, it also has implications for, say, research on irony and foreign language teaching since, for example, if a foreign language learner uses a word *w* without being aware of *w*'s semantic prosody, this may result in comical situations or, more seriously, communicating unwanted implications.

2.4 (Lexico-)Grammatical Co-occurrence: Concordances

However useful collocate displays are, for many kinds of analysis they are still not optimal. On the one hand, it is obvious that collocate displays usually provide information on lexical co-occurrence, but the number of grammatical features that is amenable to an investigation by means of collocates is limited. On the other hand, even the investigation of lexical co-occurrence by means of collocate displays can be problematic. If you investigate near synonymous adjectives such as *big, great*, and *large* (or *deadly, fatal*, and *lethal*) by looking at R1, you reduce both the precision and the recall of your results:

- precision is defined as the quotient of the number of accurate matches returned by your search divided by the number of all matches returned by your search. The collocate approach reduces precision because the R1 collocate of *big* in *the big and mean guy* is *and* rather than *guy*, and the inclusion of *and*, while of course accurate, doesn't tell you much about the semantics of *big*;[4]
- recall is defined as the number of accurate matches returned by your search divided by the number of all possible accurate matches in the data. The R1 collocate approach may reduce recall because, as we have seen in the example above, you miss *guy* in *big and mean guy* or in *the guy is big*.

The final method I am going to introduce here gets rid of this problem, though at a cost. This method, probably the most widespread corpus-linguistic tool until now, is the *concordance*. You generate a concordance if you want to know in which (larger) contexts a particular word is used. Thus, a concordance of the word *w* is a display of every occurrence of *w* together with a user-specified context; it is often referred to as KWIC (= Key Word In Context) display.[5] This user-specified context is normally either the whole sentence in which the word in question occurs (usually with some highlighting or bracketing; cf. Table 2.4.), or the word in question in a central column together with a user-specified number of words or characters to its left and its right (cf. Table 2.5 for an example). The display in Table 2.5 is especially useful because it can be sorted in various ways, such as according to the words at R1. Obviously, unlike collocate displays, concordances are read horizontally just like normal (parts of) sentences, which is in fact what they show.

While the concordance displays in Tables 2.4 and 2.5 are not always easily exploitable, they are maximally comprehensive: you can look at your search word and every possible linguistic element and how (frequently) it co-occurs with it (not just words as in the collocate display), but the price you have to pay is that you cannot usually extract this information semi-automatically as in the

Table 2.4 An example display of a concordance of *before* and *after* (sentence display)

#	Match
1	at that time and erm, we had a lot of German shutters and cameras in museum [[before]] September the third on September the fourth when I got to work they were all out.
2	It will be easy enough to bleach them with some Milk of Magnesia the night [[before]] he comes home.
3	And as, as we said [[before]], erm, many of the, erm people who lived in the poorer parts of erm the country, whether in urban or rural En erm England didn't really know about basic nutrition and and health I mean you just
4	They should be kept in an airtight jar, and rinsed in methylated spirits [[before]] wearing.
5	Yes if you looked [[after]] a child.
6	And erm some immediately post-war recipe books and I'm sure you know if you'd like to look at them [[after]] I've finished talking you might even remember some of the er er My wife still uses the.
7	This was covered by a biscuit tin lid on top of which was a kettle filled with water for an [[after]] dinner cup of tea and the washing up.
8	[[After]] the blitz on London in September nineteen forty, the government introduced a scheme whereby payments for damage to the furniture of persons earning less than four hundred pounds a year would be made, up to one hundred percent of th
9	There was a campaign [[after]] the war to bring back Tottenham puddings they had somebody in Harlow who was Well what was, what, what was Tottenham pudding then.
10	[[After]] Japan's entry into the war all imports of rubber from the far east were suspended.

Table 2.5 An example display of a concordance of *before* and *after* (tabular)

L1	Node	R1	R2
museum	before	September	the
night	before	he	comes
said	before	erm	many
spirits	before	wearing	
looked	after	a	child
them	after	I	've
an	after	dinner	cup
	After	the	blitz
campaign	after	the	war
	After	Japan	's

collocate display. In other words, concordances usually need manual analysis and annotation: In the example of *big and mean guy* from above, for example, if your corpus is not syntactically annotated, you must read the concordance line(s) yourself to determine that *big* also modifies *guy*. Given maximal explicitness, the utility of concordances is only limited by this latter fact: Inspecting a three-million-line concordance of some word is simply not feasible, and in those cases one normally retorts to heuristic techniques to filter out the relevant patterns and/or only investigate a sample of the concordance, hoping it will reflect the overall tendency well enough.

For further study/exploration:

- on corpora and corpus linguistics in general: Leech (1993); Kennedy (1998: Chapter 2); Bowker and Pearson (2002); McEnery and Wilson (2003: Chapter 2); Baker, Hardie, and McEnery (2006); and my two favorites: Biber, Conrad, and Reppen (1998) as well as McEnery, Xiao, and Tono (2006)
- on the compilation as well as the markup and annotation of corpora: Biber (1990, 1993); Sinclair (1991: Ch. 1); de Haan (1992); Leech (1993); Kennedy (1998: Chapter 2); McEnery, Xiao, and Tono (2006: Sections A.1–A.4); Beal, Corrigan, and Moisl (2007a, b), and the references cited there

- on how and why to compare corpora: Roland and Jurafsky (1998); Rayson and Garside (2000); Kilgarriff (2001); Rayson, Wilson, and Leech (2001); Gries (2007a)
- on what corpora reveal about linguistic and extralinguistic differences: Johansson (1980); Leech and Fallon (1992); Rayson, Leech, and Hodges (1997); Rayson, Wilson, and Leech (2001)
- on why frequencies matter: Ellis (2002a, b); Bod, Hay, and Jannedy (2003), in particular the paper by Jurafsky

3
An Introduction to R

But writing new scripts requires programming skills that are probably
beyond the capabilities of the average corpus linguist.

(Meyer 2002: 114 f.)

In this chapter, you will learn the foundations of R that will enable you to load, process, and store data as well as perform simple and complex text operations. Thus, this chapter prepares you for the linguistic applications in Chapter 4 and the later case study assignments.

Let us first take the most important step: the installation of R (in terms of version numbering, I am referring to R 2.8.0).

1. Download R from the CRAN website at <http://cran.at.r-project.org/>: in the section "Download and Install R", click on the link for your operating system;
    a. if you use Windows, click on base and then on the link for the setup program (i.e. <R-2.8.0-win32.exe>);
    b. if you use Linux, choose your Linux distribution (and the version of the distribution) and then the installer file (e.g., <r-base-core_2.8.0-1intrepid0_i386.deb> for Ubuntu Intrepid Ibex users, who could also use the repositories);
    c. if you use MacOS X, then choose <R-2.8.0.dmg>;
2. install all of R into the suggested standard directory;
3. start R.

That's it. You can now start and use R. However, R has more to offer. Since R is open source software, there is a very lively community of people who have written so-called packages for R. These packages are additions to R that you can load into R to obtain commands (or functions, as we will later call them) that are not part of the default configuration.

4. Enter `install.packages()`¶ at the R prompt and, when prompted for a mirror choose Austria;

5. You are then prompted to choose all packages you think you will need; if you have a broadband connection, you could choose all of them. While I will refer to some additional packages in the remainder of this book, nearly all of the code will not *require* any of them in order for you not to have to download too many packages and be at the mercy of people changing their packages. However, I will sometimes point to possibilities additional packages offer. The only package I strongly recommend getting for now is called gsubfn.

Second, you have to put some files for examples and exercises onto your hard drive. Generate a folder <C:/_qclwr/> on your hard drive (for quantitative corpus linguistics with R). Then, go to the companion website at the Google group "CorpLing with R" at <http://groups.google.com/group/corpling-with-r/web/quantitative-corpus-linguistics-with-r> and download all the files from there. Then, create three directories into which you unzip these files (the right column on the companion website tells you where to unzip which file):

- <C:/_qclwr/_assignments>; this directory then contains all larger case studies;
- <C:/_qclwr/_inputfiles>; this directory then contains all input files you require; the password you will need to unzip these files is (without the double quotes) "_hamsteR";[1]
- <C:/_qclwr/_outputfiles>; this directory then contains all output files resulting from R code in this book; the password you will need to unzip these files is (without the double quotes) "_squiRrel";
- <C:/_qclwr/_scripts>; this directory then contains all relevant R code from this book: code from explanations, examples, exercise boxes from Chapters 3 to 5 as well as all answer keys for the case study assignments; the password you will need to unzip these files is (without the double quotes) "_otteR";

Lastly, if you have a Windows-based PC, I would recommend that you now also install Tinn-R and/or Notepad++ now. Both are powerful text editors; the former is automatically optimized for writing R code:

1. Download Tinn-R from the SciViews website at <http://www.sciviews.org/Tinn-R/>;
2. Double-click on <Tinn-R_*_setup.exe>;
3. Install the program with the default settings into the suggested standard directory; make sure you tick the option that makes Tinn-R the default editor for .r files;
4. You can then start Tinn-R by double-clicking on the icon on the desktop, the icon in the start menu, or the icon in the quick launch tool bar. Make sure that Tinn-R is set to syntax highlighting for R: go to "*Options: Syntax: Set: R*" or, better even, set R as default by going to "*Options: Syntax: Default: R*".

Notepad++ (cf. <http://notepad-plus.sourceforge.net/uk/site.htm>) is not optimized for writing R code right from the start, but its syntax highlighting can be configured very easily in the right way and it is more flexible when it comes to handling Unicode characters. As a result of this, you will be able to look at all the scripts in <C:/_qclwr/_scripts> with Tinn-R or Notepad++, which will make it easier for you to understand them. If you cannot use Tinn-R or Notepad++, try JGR (cf. <http://rosuda.org/JGR/>) or the R-internal editor, which you can access via the menu "*File: New Script*" or "*File: Open Script*", but which has no syntax highlighting. As a Mac user, you can use the R-internal editor, too, or R.app (cf. <http://cran.r-project.org/bin/macosx/RMacOSX-FAQ.html>). As a Linux user, you can also use Emacs (cf. <http://www.gnu.org/software/emacs/> or <http://ess.r-project.org/>) or just gedit (which comes with Ubuntu and has R syntax highlighting); cf. the

Appendix for links to other editors. Let me also recommend that you download from the URLs in the Appendix the R reference card (the short or the long version) and/or the R/Rpad reference card so that apart from this book and your own notes you also have a quick cheatsheet available.

Now, my recommendations for self-study are the following:

- up to Section 3.6: read the book at your computer and enter the lines from the book directly into R and execute them;
- for Section 3.6 and following as well as Chapters 4 and 5: read the book at your computer and enter the lines from the book either directly into the R console or into Tinn-R (or some other editor) and then copy and paste them into R;
- for the case study assignments from the companion website: read the assignments and try to solve them using Tinn-R or some other editor and OpenOffice.org Calc; then check up on your solutions with the answer key provided in my script and output files.

Basically, it is best if you write all code that is longer than two to three lines or very complex in Tinn-R and then copy and paste it into R, because the syntax highlighting function of Tinn-R makes it much easier to avoid mistakes or, if you still made a mistake, find and correct it. This means, for most of Chapter 3, you will not need Tinn-R, but by the time you begin with Chapter 4, you should always work in Tinn-R or R's editor as suggested above.

As was already mentioned in the introduction, R is an extremely versatile piece of software. It can be used as a calculator, a spreadsheet program, a statistics program, a (statistical) graphics program, and a scripting programming language with sophisticated mathematical and character processing capabilities (both for ASCII and Unicode).[2] In fact, the range of functions R can perform is so vast that I will have to restrict my discussion to a radically reduced subset of its functions. The functions that will be covered in this book are mainly concerned with

- how to load, process, and save various kinds of data structures with a special emphasis on handling text data from corpora for corpus-linguistic analyses;
- how to load, process, and evaluate tabular data (for statistical/graphical analysis).

Thus, I will unfortunately have to leave aside many intriguing aspects of R. One of these, which I would like you to explore immediately when you begin to write your own R code, is how to write your own functions. For reasons of space this cannot be discussed here but it is one of the most powerful aspects of R (or other programming languages such as Perl or Python). As a first step, when you have finished the book or at least Chapter 4, you might want to read up on this in the documentation that comes with R; cf. below. It's worth it! Also, for didactic and expository reasons I will sometimes not present the shortest or most elegant way of handling a particular problem but rather present a procedure that is more adequate, for one or more reasons. These reasons include the desires

- to keep the number of functions, arguments, and regular expressions manageable;
- to highlight similarities between different functions;
- to allow you to recycle parts of code.

Thus, this book is not a general introduction to R, and while I aim at enabling you to perform a multitude of tasks with R, I advise you to also consult the additional references in the reference section and the comprehensive documentation that comes with R, and I will mention below how you can access these help files.

First, some notational conventions. URLs, files, and folders will be mentioned like this: <C:/_qclwr/_inputfiles/some_name.txt>, where "some_name" will refer either to a data source or the chapter, section, and sub-section in which the data will be used. Input to R is usually given in blocks of code like this, where "•" means a space, and "¶" denotes a line break (i.e. an instruction for you to press ENTER):

```
>•mean(c(1,•2,•3))¶
[1]•2
```

This means for you: do *not* enter the two characters ">•", i.e. smaller-than followed by space. These are only provided for you so that (i) you know that this is code you are supposed to enter—don't enter grey-shaded code that does not begin with ">•")—and (ii) you can distinguish your *in*put from R's *out*put more easily. You will also occasionally see lines that begin with "+" or "+•". These plus signs (and spaces), which you are *not* supposed to enter either, begin lines where R is still expecting further input from you before it begins to execute the function. For example, when you enter "2-" and press ENTER, then this is how your R interface will look like:

```
>•2-¶
+
```

R is waiting for you to complete the subtraction. When you enter the number you wish to subtract and press ENTER, then the function will be executed properly.

```
+•3¶
[1]•-1
```

Another example: if you wish to load the package "corpora" into R to access some of the functions that the corpus linguist Stefan Evert has contributed to the community, you can load this package by typing `library(corpora)`¶. (Note: this only works when you have installed the package as explained above.) However, if you forget the closing bracket, R will wait for you to complete the input:

```
>•library(corpora¶
+•)¶
>
```

Unfortunately, R will not always be so forgiving . . . By the way, if you make a mistake in R, you often need to change only one minor thing in a line. Thus, rather than typing the whole line again, just press the cursor-up key as many times as is necessary to get back to the line you wish to change or execute again.

Corpus files or tables/data frames will be represented as in Figure 3.1, where "→" and "¶" again denote tab stops and line breaks respectively.

PartOfSpeech	→	TokenFrequency	→	TypeFrequency	→	Class ¶
ADJ	→	421	→	271	→	open ¶
ADV	→	337	→	103	→	open ¶
N	→	1411	→	735	→	open ¶
CONJ	→	458	→	18	→	closed ¶
PREP	→	455	→	37	→	closed ¶

Figure 3.1 Representational format of corpus files and data frames.

Menus, submenus, and commands in submenus in applications are given in italics in double quotes, and hierarchical levels within application menus are indicated with colons. So, if you open a document in, say, Microsoft Word or OpenOffice.org Writer, you do that with what is given here as "*File: Open...*"

Before we delve into R, let me finally mention several ways of using R's own documentation. First, as a Windows user, if you want to read about R, start R, click on "*Help: HTML help,*" then on "An introduction to R" or "Writing your own functions." Second, if you know the name of a function or construct and just need some information about it, type a question mark directly followed by the name of the function at the R prompt to access the help file for this function (e.g. ?help¶, ?sqrt¶, ?"["¶, or ...). Third, if you know what you would like to do but don't know the function, start R, click on "*Help: HTML help,*" then on "Search Engine and Keywords." In the window, you will then see that you can either browse the keywords by topic or enter words to search the index of the R HTML help files. Alternatively, you can either enter help.start()¶ or help .search("...")¶ and instead of "..." you enter what you're interested in (e.g. "correlation" or "replace" or ...).

3.1 A Few Central Notions: Data Structures, Functions, and Arguments

The various functions that are going to be explored in the following sections concerning R will be technical and relatively abstract. While I will often make use of linguistic examples to explain and exemplify a particular point, the true corpus-linguistic relevance and power of the methods to be introduced will probably only transpire a little later.

The simplest way of using R is using it like you would use a pocket calculator. At the R prompt, you just enter any arithmetic operation and hit ENTER:[3]

```
>•2+2¶
[1]•4
>•3^2¶
[1]•9
>•(2+3)^2¶
[1]•25
```

As you may recollect from your math classes in school, however, you often had to work with variables rather than the numbers themselves. R can also operate with variable names, where the names represent the content of so-called data structures. Data structures in R can take on a variety of forms. The simplest one, a vector, can contain just a number; the most complex one, a list, can contain many large tables, texts, numbers, and other data structures. Data structures can be entered into R at the prompt, but the more complex a data structure becomes, the more likely it of course becomes that you read it from a file, and this is in fact what you will do most often in this book: reading in tables and text files, processing text files, performing computations on tables, and saving texts or tabular outputs into text files.

One of the most central things to understand about R is how you tell it to do something other than the simple calculations from above. A command in R virtually always consists of two elements: a *function* and, in parentheses, *arguments*, where arguments can be null, in which cases there are just opening and closing parentheses. The function is an instruction to do something, and the arguments to a function represent (i) what the instruction is to be applied to and (ii) how the instruction is to be applied to it. Let us look at two simple arithmetic functions you know from school. If you want to compute the square root of 5 with R—without simply entering the instruction 5^0.5¶, that is—you need to know the name of the function and how many and which arguments it takes. Well, the name of the function is `sqrt`, and it takes just one argument, namely the number of which you want the square root:

```
>•sqrt(5)¶
[1]•2.236068
```

As another example, let us compute the base-10 logarithm of 150. One possibility would be to use `log10`, which, just like `sqrt`, only takes one argument, namely the number of which you want the logarithm. However, since one may also want logarithms to other bases, there is a more flexible function in R, which takes two arguments: `log`. The first is the number of which you want the logarithm, the second is the base to the logarithm. Thus, you can enter this:

```
>•log(150,•10)¶
[1]•2.176091
```

One important aspect here is that this way of computing the log is only a short version of the more verbose version:

```
>•log(x=150,•base=10)¶
[1]•2.176091
```

In this longer version, the arguments provided are labeled: `x` for the number of which you want the logarithm and `base` for the base. If, however, you provide arguments in *exactly* the order in which R expects them, you can leave the labels out—if you change the order or want to leave out one argument, then you must label the arguments so that R knows what in your list of arguments means what.

Three other things worth mentioning. First, outside of quotes, R does not care whether you put a space before the comma or not. Second, R's understanding of its input is case-sensitive. Third, as you see below, R ignores everything after a "#", so you can use this to comment your lines when you write small scripts and want to tell yourself what a particular line is doing.

```
>•Log(150,•10)•#•R•does•not•know•this•function•-•it•only•knows•"log"¶
Error:•couldn't•find•function•"Log"
```

Note that, in all these cases, R does not store the result—it just outputs it. When you want to assign some content to some data structure for later use, you must use the assignment operator "<-" (a less-than sign followed by a minus). You can basically choose any name as long as it contains only letters, numbers, underscores, or periods and starts with a letter or a period. However, to avoid confusion, you should not use names for data structures that are R functions (such as log or sqrt or the many other functions you'll get to know below). As a result of assignment, the content resulting from the function is available in the data structure just defined. For example, this is how you store the square root of 5 into the kind of data structure you will get to know as a vector.

```
>•aa<-sqrt(5)•#•compute•the•square•root•of•5¶
```

You can now check whether R actually knows about this vector now by instructing R to list all the data structures it knows about (in the current R session):

```
>•ls()¶
[1]•"aa"
```

And you can instruct R to output the contents of the vector (by default to the screen):

```
>•aa¶
[1]•2.236068
```

A short way to simultaneously instruct R to assign a value to a data structure *and* output that data structure is by putting the assignment into parentheses:

```
>•(aa<-sqrt(5))¶
[1]•2.236068
```

Note also that the assignment operator can also be used to change the value of an existing data structure using the name of the data structure again.

```
>•(aa<-aa+2)¶
[1]•4.23606
```

By the way, you can enter more than one command per line by separating commands with a semicolon:

```
>•sqrt(9);•sqrt(16)¶
[1]•3
[1]•4
```

If you ever want to get rid of data structures again, you can also remove them. Either you remove an individual data structure by using rm (for remove) and provide the data structure to be deleted as an argument:

```
>•rm(aa)•#•delete•aa¶
```

or you can just delete all data structures.

```
>•rm(list=ls(all=TRUE))•#•delete•all•data•structures¶
```

Finally, before we look at several data structures in more detail shortly, note that not only may functions not require their arguments to be labeled, many functions even have default settings for arguments which they use when the argument is not called upon explicitly by the user at all. A function to exemplify this characteristic that is actually also very useful in a variety of respects is sample. This function outputs random or pseudo-random samples, which is useful when, for example, you have a large number of matches or a large table and want to access only a small, randomly chosen sample. The function sample can take up to four arguments:

```
sample(x,•size,•replace=FALSE,•prob=NULL)•#•no•">•"•at•the•beginning:•
  don't•enter•this
```

- x: a data structure, most likely a vector, providing the elements from which you want a sample. If that data structure x is a vector of one number only, then R outputs a sample of the numbers from 1 to x; if x is a vector of two or more elements, then R outputs a sample of the elements of that vector.
- size: an integer number determining the size of the sample.
- a logical expression: replace=TRUE or, shorter, replace=T (if the elements of x can be sampled multiple times because they are, as it were, replaced into the bin after having been sampled) or replace=FALSE or, shorter, replace=F (if the elements of x can only be chosen once because they are not replaced into the bin, the default setting).
- prob: a vector of probabilities with which elements of x may be sampled; the default setting is NULL, representing equiprobable sampling, i.e. sampling where each element of x has the same chance of being sampled.

Let us look at a few examples that make successively more use of label omission and default settings. First we generate a vector with the numbers of 1 to 10:

```
>•x<-c(1:10)¶
```

If you now want to draw five elements equiprobably and randomly out of this sample with replacement, you enter this:[4]

```
>•sample(x,•size=5,•replace=T,•prob=NULL)¶
[1]•5•9•9•9•2
```

However, since you provide the arguments in the default order, you can do away with the labels (although you now of course get different random numbers):

```
>•sample(x,•5,•T,•NULL)¶
[1]•3•8•4•1•7
```

But since prob=NULL is the default setting, you might as well omit it altogether:

```
>•sample(x,•5,•T)¶
[1]••2••1••9••9••10
```

And this is how we draw five elements equiprobably and randomly *without* replacement:

```
>•sample(x,•5,•F)¶
[1]••1•10••6••3••8
```

But again, since replace=F is the default, why not omit it?

```
>•sample(x,•5)¶
[1]•10••5••9••3••6
```

You have now seen that the arguments prob and replace can be omitted because of their default settings. But actually, it is also possible to omit the size argument. If the size argument is omitted, R assumes we want all elements of the vector x back again and, thus, effectively only provides us with a random order of the elements of x:

```
>•x¶
[1]••1••2••3••4••5••6••7••8••9•10
>•sample(x)¶
[1]••2••4••3•10••9••8••1••6••5••7
```

In fact, if the idea is just to get a random ordering of the numbers of 1 to *x*, then you do not even need to provide R with a vector giving all the numbers from 1 to 10, as you did when we defined x (cf. above). You can just give R the number:

```
>•sample(10)¶
[1]••5•10••2••6••1••3••4••9••7••8
```

In fact, default settings sometimes lead to the extreme case of function calls without *any* argument (like help.start() above). The function q shuts R down and internally processes three arguments:

- a character string—i.e. a sequence of characters such as letters, numbers, or other symbols—specifying whether the current workspace, i.e. the data structures you worked with since you last started R, should be saved or the user should be prompted to specify whether the workspace should be saved; the latter is the default.
- a number specifying what R should "tell" the operating system when R is shut down; the default is 0, which means "successful shut down".
- a logical value (T or F) specifying whether some other user-specified function should be executed before R shuts down: the default is T.

Thus, if you want to shut down R, you just enter:

```
>•q()¶
```

Since you have not provided any arguments at all, R assumes the default settings. The first requires R to ask you whether the workspace should be saved. Once you answer this question, R will shut down and send 0 (i.e. "successful shut down") to your operating system.

As you can see, label omission and default settings can be very useful ways of minimizing typing effort. However, especially at the beginning, it is probably wise to try to strike a balance between minimizing typing on the one hand and maximizing transparency of the code on the other hand. While this may ultimately boil down to a matter of personal preferences, I recommend using more explicit code at the beginning in order to be maximally aware of the options your R code uses. The next sections will introduce data structures that are most relevant to linguistic and statistical analysis.

3.2 Vectors

3.2.1 Basics

The most basic data structure in R is a vector. Vectors are one-dimensional, sequentially ordered sequences of elements (such as numbers or character strings (such as words)). While it may not be completely obvious why vectors are important here, we must deal with them in some detail since nearly all other data structures in R can ultimately be understood in terms of vectors. As a matter of fact, you have already used vectors when we computed the square root of 5:

```
>•sqrt(5)¶
[1]•2.236068
```

The "[1]" in front of the result indicates that R internally represents the result as a vector and that the number that is printed is the first (and here also only) number of that vector.[5] You can even test this with R itself. If you assign the data structure resulting from sqrt(5)¶ to aa and then ask R whether aa is a vector, we get R's version of "yes":

```
>•aa<-sqrt(5)•#•compute•the•square•root•of•5¶
>•is.vector(aa)¶
[1]•TRUE
```

You can also find out what kind of vector aa is using the function class, and you can also determine that the length of aa is 1:

```
>•class(aa)¶
[1]•"numeric"
>•length(aa)¶
[1]•1
```

And, you can define 'empty' vectors with a particular length, which may seem senseless right now, but is still something you should remember well because it is more memory-efficient than just letting a vector grow dynamically; this will become clearer below.

```
>•(empty<-vector(length=3))¶
[1]•FALSE•FALSE•FALSE
```

For corpus linguists, it is of course important to know that vectors can also contain character strings—the only difference with numbers is that the character strings have to be put either between double quotes or between single quotes. You can freely choose which kind of quotes you use, but the opening and the closing quote must be identical. (For the sake of consistency, I will only use double quotes in the whole book.)

```
>•(a.name<-"James")¶
[1]•"James"
>•class(a.name)¶
[1]•"character"
```

Note what happens when you apply length to a.name. Contrary to what you might initially expect, you do *not* get the number of characters of "James" (i.e. 5)—you get the length of the data structure, and since the vector contains one element, "James", R returns 1:

```
>•length(a.name)¶
[1]•1
```

Actually, there are six different types of vectors, but we will restrict our attention to vectors of real numbers and vectors of character strings. However, vectors usually only become interesting when they contain more than one item. The function that concatenates (i.e. combines) several elements into a vector is called c, and its arguments are the elements to be concatenated into a vector.

```
>•(numbers<-c(1,•2,•3))¶
[1]•1•2•3
>•(names<-c("James",•"Jonathan",•"Jean-Luc"))¶
[1]•"James"••••"Jonathan"•"Jean-Luc"
```

Since an individual number such as the square root of 5 is already a vector, it is not surprising that c also connects vectors consisting of more than one element (as does append):

```
>•numbers1<-c(1,•2,•3);•numbers2<-c(4,•5,•6)•#•generate•two•vectors¶
>•(numbers1.and.numbers2<-c(numbers1,•numbers2))•#•join•  the•two•vectors¶
[1]•1•2•3•4•5•6
>•(numbers1.and.numbers2<-append(numbers1,•numbers2))•#•another•
  way•to•join•the•two•vectors¶
[1]•1•2•3•4•5•6
```

A characteristic that will be useful further below is that R can handle and apply vectors recursively. For example, adding two equally long numerical vectors yields a vector with all pairwise sums:

```
>•numbers1+numbers2•#•that•is,•1+4,•2+5,•3+6¶
[1]•5•7•9
```

What happens if vectors are not equally long? Two things can happen. First, if the length of the longer vector is divisible without a remainder by the length of the shorter vector—i.e. if the modulus of the two lengths is 0—then the shorter vector is recycled as often as necessary to complete the function. The most frequent such case is when the shorter vector has the length 1:

```
>•bb<-10¶
>•numbers1*bb¶
[1]•10•20•30
```

Second, if the length of the longer vector is *not* divisible without a remainder by the length of the shorter vector, the operation proceeds as far as possible, but also returns a warning:

```
>•bb<-c(10,•20)¶
>•numbers1*bb¶
[1]•10•40•30
Warning•message:
In•numbers1•*•bb•:
••longer•object•length•is•not•a•multiple•of•shorter•object•length
```

Another characteristic you will use a lot later is that elements of vectors can be named:

```
>•names(numbers)<-c("first",•"second",•"third");•numbers¶
•first•second••third
••••1••••••2••••••3
```

It is important to note that—unlike arrays in Perl—vectors can only store elements of one data type. For example, a vector can contain numbers *or* character strings, but not really both: if you try to force character strings into a vector previously containing only numbers, R will change the data type, and since you can interpret numbers as characters but not vice versa, R changes the numbers into character strings and then concatenates them into a vector of character strings:

```
>•(mixture<-c(1,•2,•"Benjamin"))¶
[1]•"1"••••••••"2"••••••••"Benjamin"
```

The double quotes around the 1 and the 2 indicate that these are now understood as character strings, which also means that you cannot use them for calculations any more (unless you change their data type back using as.numeric). Apart from class, you can identify the type of a vector (or the data types of other data structures) with str (for 'structure'), which takes as an argument the name of a data structure:

```
>•str(numbers1)¶
num•[1:3]•1•2•3
>•str(mixture)¶
chr•[1:3]•"1"•"2"•"Benjamin"
```

Unsurprisingly, the first vector consists of numbers, the second one of character strings. In most cases to be discussed here, vectors will not be entered into R at the console but will be read in from files. In the following section, we will be concerned with how to load files containing text.

For further study/exploration:

- on how to repeat a vector a user-specified number of times: ?rep¶
- on how to generate user-defined regular sequences: ?seq¶
- on how to change the types of vector to numeric and character strings: ?as.numeric¶ and ?as.character¶

3.2.2 Loading Vectors

R has a very powerful function to load the contents of text files into vectors, scan. Since this function is central to just about all corpus-loading operations to be discussed below, we will discuss it and a variety of its arguments in quite some detail. The most important arguments of scan for our purposes together with their default settings are as follows:

```
scan(file="",•what=double(0),•sep="",•quote=if(identical(sep,•"\n"))•
  ""•else•"'\"",•dec=".",•skip=0,•quiet=F,•  comment.char="")¶
```

- The file argument is obligatory (for loading vectors from files) and specifies the path to the file to be loaded; usually this will look like this "E:/Corpora/BNCwe/d8y.txt". If you have data in the clipboard that you want to load, you can also write file="clipboard". If you use MacOS X or Linux, you can use file=file.choose(), and on Windows you can also use the most convenient argument file=choose.files().[6]
- The what argument specifies the kind of input scan is supposed to read. The most important settings are what="double(0)" and what="character" (or an abbreviation of this, such as "char"): you use the former if the file contains only numbers (but you need not because that's the default anyway) and the latter if the file contains text.
- The sep argument specifies the character that separates individual entries in the file. The default setting, sep="", means that any whitespace character will separate entries, i.e. spaces, tabs (represented as "\t"), and newlines (represented as "\n" but cf. below). Thus, if you want to read in a text file into a vector such that each line is one element of the vector, you write sep="\n", which is what we will do most of the time.
- The quote argument specifies which characters surround text quotes, most of the time sep will be set as sep="\n", which entails that quote is then automatically set to quote="".
- The dec argument specifies the decimal point character; if you want to use a comma instead of a period, just enter that here as dec=",".
- The skip argument specifies the number of lines you wish to skip when reading in a file, which may be useful when, for example, corpus files have a fixed number of header rows at the beginning of the file.
- The quiet argument specifies whether R returns the number of entries it has read in (quiet=F) or not (quiet=T).

Let us look at a few examples. First, we load into a vector x the contents of a text file <C:/_qclwr/_inputfiles/dat_vector-a.txt> that looks like Figure 3.2.

```
1¶

2¶

3¶
```

Figure 3.2 The contents of <C:/_qclwr/_inputfiles/dat_vector-a.txt>.

You just enter the following line (or use one of the above alternatives), making use of the fact that R supplies the default settings unless you specify them explicitly:

```
>•x<-scan(file="C:/_qclwr/_inputfiles/dat_vector-a.txt",•  sep="\n")¶
Read•3•items
```

As a result, x is now stored in R's working memory and can be printed out onto the screen. A slightly more complex example: imagine you have a file <C:/_qclwr/_inputfiles/dat_vector-b.txt> looking like Figure 3.3.

```
This·is·the·first·line¶
This·is·the·second·line¶
```

Figure 3.3 The contents of <C:/_qclwr/_inputfiles/dat_vector-b.txt>.

These are two ways to load this file into a vector x. (Note again that you can abbreviate "character".) Before you read on, can you guess what the difference is between them both?

```
>•x.1<-scan(file="C:/_qclwr/_inputfiles/dat_vector-b.txt",•what="char")¶
>•x.2<-scan(file="C:/_qclwr/_inputfiles/dat_vector-b.txt",•.what="char",•
  sep="\n")¶
```

THINK

BREAK

The first line reads in the file with the default setting of sep, so everything separated by spaces in the original file becomes an element in the vector x.1. The second line reads in the file such that everything separated by line breaks becomes an element in the vector x.2:

```
>•x.1¶
[1]•"This"•••"is"•••••"the"••••"first"••"line"•••"This"•••"is"•••••"the"••••"
   second"•"line"
>•x.2¶
[1]•"This•is•the•first•line"••"This•is•the•second•line"
```

As I mentioned above, R for Windows has a particularly attractive function, which allows you to choose one or even several existing files interactively in an explorer window without having to enter long and complex paths: `choose.files()`. On MacOS or Linux you have to proceed differently: if you want to choose *one* already existing file, you can use `file.choose()`, and since this function works on all operating systems, this is what I will use here whenever you are supposed to choose one existing file.

If you want to choose *several* already existing files on MacOS or Linux, you can use the following in place of `choose.files()`, which will prompt you to enter a directory path and then either open a window of the operating system or, on Linux, ask you to choose files by entering their numbers in a list that R will provide.

```
>•filenames<-select.list(dir(scan(nmax=1,•what="char")),•multiple=T)¶
```

Thus, instead of the above paths for x.1 and x.2, you could also enter the following to be prompted to choose a file, which R would then read as a vector of character strings:

```
>•x.1<-scan(choose.files(),•what="char")¶
>•x.2<-scan(choose.files(),•what="char",•sep="\n")¶
```

An application relevant to corpus-linguistic studies is that you may have a file—let's call it <my_corpusfiles.txt>—with the paths to all corpus files you may wish to search. This file may look like Figure 3.4:

```
C:/temp/corpusfile_01.txt¶
C:/temp/corpusfile_02.txt¶
C:/temp/corpusfile_03.txt¶
```

Figure 3.4 The contents of a fictitious(!) file <C:/_qclwr/my_corpusfiles.txt>.

and this is how you tell R what your corpus files are:

```
corpus.files<-scan(choose.files(),•what="char",•sep="\n")
```

You would be prompted in an explorer window to choose <C:/_qclwr/my_corpusfiles.txt>, which would then be loaded into the vector corpus.files. If you wanted to input more than one file name and, because you work on MacOS or Linux, cannot use choose.files(), you could use the above option with select.list or dir, which I will talk about in Section 3.8.

Finally, while loading files will be the most frequent way in which you will use scan, let me mention the simplest way in which you can use scan, namely to enter vectors. If you just write scan()¶ or scan(what="char")¶, you can enter numbers or strings separated by ENTER until you press ENTER twice to end your input:

```
>•x<-scan()¶
1:•1¶
2:•2¶
3:•3¶
4:¶¶
Read•3•items
>•x¶
[1]•1•2•3
```

For further study/exploration:

- on how to load a file linewise: ?readLines¶ and Spector (2008: Section 2.7)
- on how to customize input with scan: ?scan¶

3.2.3 Accessing and Processing (Parts of) Vectors

Now that we have covered some aspects of how to load vectors, let us turn to how to access parts of vectors and do something with these parts. Some elementary ways of accessing specific parts of vectors of numbers are provided by the functions min and max, which take as their arguments a vector of numbers and output the minimum and maximum value respectively:

```
>•min(c(1,•2,•3));•max(c(1,•2,•3))¶
[1]•1
[1]•3
```

But these functions are restricted to numerical vectors only and not particularly flexible. The simplest ways to get a glimpse of what any data structure looks like are the functions head and tail. These take the name of a data structure as one argument and return the six first (for head) or last (for tail) elements of a data structure; if you want a number of elements other than 6, you can provide that as a second argument to head and tail. However, one of the most powerful ways to access parts of data structures is by means of subsetting, i.e. indexing with square brackets. In its simplest form, you can access (and of course change) a single element:

```
>•x<-c("a",•"b",•"c",•"d",•"e")¶
>•x[3]•#•access•the•3rd•element•of•x¶
[1]•"c"
```

Since we have already seen how flexibly R handles variables, the following extensions of this simple principle should not come as a surprise.[7]

```
>•y<-3¶
>•x[y]•#•access•the•3rd•element•of•x¶
[1]•"c"
>•z<-c(1,•3)¶
>•x[z]•#•access•the•1st•and•the•3rd•element•of•x•just•as•x[c(1,•3)]•would•do¶
[1]•"a"•"c"
>•z<-c(1:3)¶
>•x[z]•#•access•the•elements•1•to•3•of•x¶
[1]•"a"•"b"•"c"
```

With negative numbers, you choose the data structure *without* the designated elements:

```
>•x[-2]•#•access•x•but•without•its•2nd•element¶
[1]•"a"•"c"•"d"•"e"
```

R also offers more useful functions, however. One of the most interesting ones is to let R decide which elements of a vector satisfy a particular condition. The simplest though not always most elegant solution is to present R with a logical expression.

```
>•x=="d"¶
[1]•FALSE•FALSE•FALSE••TRUE•FALSE
```

R checks all elements and outputs for each whether it meets the condition given in the logical expression or not. The one thing you need to bear in mind is that the logical expression uses "= =" rather than the "=" we know from the assignment of values to arguments. Other logical operators we will use are the following:

&	and	|	or
>	greater than	>=	greater than or equal to
<	less than	<=	less than or equal to
!	not	!=	not equal

The following examples illustrate these logical expressions:

```
>•(x<-c(10:1))•#•generate•and•output•a•vector•with•the•numbers•from•10•to•1¶
[1]•10••9••8••7••6••5••4••3••2••1
>•x==4•#•which•elements•of•x•are•4?¶
[1]•FALSE•FALSE•FALSE•FALSE•FALSE•FALSE••TRUE•FALSE•FALSE•FALSE
>•x<=7•#•which•elements•of•x•are•smaller•than•or•equal•to•7?¶
[1]•FALSE•FALSE•FALSE••TRUE••TRUE••TRUE••TRUE••TRUE••TRUE••TRUE
>•x!=8•#•which•elements•of•x•are•not•8?¶
[1]••TRUE••TRUE•FALSE••TRUE••TRUE••TRUE••TRUE••TRUE••TRUE••TRUE
>•(x>8•|•x<3)•#•which•elements•of•x•are•larger•than•8•or•smaller•than•3?¶
[1]••TRUE••TRUE•FALSE•FALSE•FALSE•FALSE•FALSE•FALSE••TRUE••TRUE
```

While this illustrates how logical expressions work, it probably also illustrates that this is not a particularly elegant way to proceed. Regardless of how many elements of x fulfill the condition, you always get ten truth values and must then identify the cases of TRUE yourself. Thankfully, there is a more elegant way of doing this using which, which requires as an argument a logical expression and outputs the position of the element(s) satisfying a particular condition:

```
>•which(x==4)•#•which•elements•of•x•are•4?¶
[1]•7
```

The following examples correspond to the ones we just looked at but use which:

```
>•which(x<=7)•#•which•elements•of•x•are•less•than•or•equal•to•7?¶
[1]••4••5••6••7••8••9•10
>•which(x!=8)•#•which•elements•of•x•are•not•8?¶
[1]••1••2••4••5••6••7••8••9•10
>•which(x>8•|•x<3)•#•which•elements•of•x•are•larger•than•8•or•less•than•3?¶
[1]••1••2••9•10
```

A central point here is not to mix up the *position of an element* in a vector and the *element* in a vector. If we come back to which(x==4)¶, remember that it does not output 4—the element—but 7—the position that 4 occupies in x. If, however, you have been following along so far, you may already be able to guess how you get at the element so try to write a line that retrieves the elements of x that are larger than 8 or smaller than 3 and stores them in a vector y.

THINK

BREAK

Since you can access elements of a vector with square brackets and since the output of which is itself a vector, you can simply write:

```
>•(pointer<-which(x>8•|•x<3))¶
[1]••1••2••9•10
>•(y<-x[pointer])¶
[1]•10••9••2••1
```

Or, because you guess that you can combine it all into one expression:

```
>•x[which(x>8•|•x<3)]•#•output•the•elements•of•x•which•are•greater•than•8•
   or•smaller•than•3¶
[1]•10••9••2••1
```

or even:

```
>•x[x>8•|•x<3]•#•output•the•elements•of•x•which•are•greater•than•8•
   or•smaller•than•3¶
[1]•10••9••2••1
```

You can also apply length to find how many elements of a vector fulfill this condition:

```
>•length(which(x>8•|•x<3))•#•output•the•number•of•elements•of•x•
   which•are•greater•than•8•or•smaller•than•3¶
[1]•4
```

Or, since TRUE and FALSE are taken as 1 and 0 respectively, you can write an even shorter version:

```
>•sum(x>8•|•x<3)•#•output•the•number•of•elements•of•x•which•are•
   greater•than•8•or•smaller•than•3¶
[1]•4
```

Thus, the fact that R uses vectors for nearly everything is in fact a great strength and makes it a very versatile language. When you combine the above with subsetting, you can also change elements of vectors very quickly. For example, if you want to replace all elements of x that are greater than 8 by 12, this is one way of achieving it:

```
>•x•#•output•x¶
[1]•10••9••8••7••6••5••4••3••2••1
>•y<-which(x>8)•#•store•the•positions•of•elements•greater•than•9•in•the•
   vector•y¶
>•x[y]<-12;•x•#•replace•the•elements•of•x•that•are•greater•than•8•by•12¶
[1]•12•12••8••7••6••5••4••3••2••1
```

or even:

```
>•x<-c(10:1)•#•generate•a•vector•with•the•numbers•from•10•to•1¶
>•x[which(x>8)]<-12;•x•#•change•the•element(s)•in•x•which•are•greater•
   than•8•to•12¶
[1]•12•12••8••7••6••5••4••3••2••1
```

or even:

```
>•x<-c(10:1)•#•generate•a•vector•with•the•numbers•from•10•to•1¶
>•x[x>8]<-12;•x¶
[1]•12•12••8••7••6••5••4••3••2••1
```

Apart from which and subsetting, there are also some more powerful functions available for processing vectors. We will briefly look at %in% and match, but discuss only the simplest cases where they take two arguments. The first argument of %in% is a data structure (usually a vector) with the elements to be matched; the second is a data structure (typically a vector or a data frame) with the elements to be matched against. The output is a logical vector of the length of the first argument with TRUEs and FALSEs for the elements of the first data structure that are found and that are not found in the second data structure respectively. Let me clarify this:

```
>•x<-c(10:1);•y<-c(2,•5,•9)•#•generate•vectors•again¶
>•x•%in%•y¶
[1]•FALSE••TRUE•FALSE•FALSE•FALSE••TRUE•FALSE•FALSE••TRUE•FALSE
>•y•%in%•x¶
[1]•TRUE•TRUE•TRUE
```

The first example, x•%in%•y¶, shows that the first element of x—10—does not appear in y, while the second element—9—does, etc. You can combine this with subsetting to get at the matched elements quickly:

```
>•x[x•%in%•y]¶
[1]•9•5•2
```

The function match returns a vector of the positions of (first!) matches of its initial argument in its second (again, both arguments are typically vectors):

```
>•match(x,•y)¶
[1]•NA••3•NA•NA•NA••2•NA•NA••1•NA
```

This tells you, as above, that the first element of x does not appear in y, but that the second element of x—9—is the third element in y. The third, fourth, and fifth element of x do not appear in y, but the sixth element—5—appears in y, namely at position 2 of y, etc. It should now be clear what the following line does, although it is sometimes a little tricky to see:

```
>•match(y,•x)¶
[1]•9•5•2
```

This means: the first element of y—the 2—is the ninth element of x. The second element of y—the 5—is the sixth element of x. And so on. You may ask yourself now, what this is good for, but you will get to see very useful applications of these two functions below.

> For further study/exploration:
>
> • on how to identify duplicated elements in a vector, which can be useful for type-token counts: the number of types is the number of all non-duplicated tokens: ?duplicated¶
> • on how to seek partial matches for the elements of the first argument among those of the second: ?pmatch¶ and ?charmatch¶
> • Hans-Jörg Bibiko's function levenshtein can be used to find minimal pairs (to be found at <http://wiki.r-project.org/rwiki/doku.php?id=tips:data-strings:levenshtein>)

Another related way of processing two vectors uses basic set-theoretic concepts. Let me introduce three such functions—setdiff, intersect, and union, which all take two vectors as arguments—on the basis of the two vectors x and y that we just used. The function setdiff returns the elements of the vector given as the first argument that are not in the vector given as the second argument:

```
>•setdiff(x,•y)¶
[1]•10••8••7••6••4••3••1
>•setdiff(y,•x)¶
numeric(0)
```

The function intersect provides the elements of the vector of the first argument that also occur in the vector given as the second argument; note that the order in which the elements are returned is determined by the order in which they occur in the vector given as the first argument.

```
>•intersect(x,•y)¶
[1]•9•5•2
>•intersect(y,•x)¶
[1]•2•5•9
```

The function union provides the elements that occur at least once in the combination of the two vectors given as arguments; again, the order depends on which vector is listed first:

```
>•union(x,•y)¶
[1]•10••9••8••7••6••5••4••3••2••1
>•union(y,•x)¶
[1]••2••5••9•10••8••7••6••4••3••1
```

Let us now briefly also mention the very useful functions `unique` and `table`. The function `unique` takes as its argument a vector or a factor and should be especially easy to explain to linguists since it outputs a table containing all the types that occur at least once among the tokens (i.e. elements) of said vector/factor:

```
>•g<-c(1,•2,•3,•2,•3,•4,•3,•4,•5)¶
>•h<-c(2,•3,•1,•5,•2,•6,•3,•1,•2)¶
>•unique(g)¶
[1]•1•2•3•4•5
```

The function `table` takes as argument one or more vectors or factors and provides the token frequency of each type or each combination of types:

```
>•table(g)¶
g
1•2•3•4•5
1•2•3•2•1
```

That is, in g there is one 1, there are two 2s, etc. Now what happens with two vectors?

```
>•table(g,•h)¶
•••h
g•••1•2•3•5•6
••1•0•1•0•0•0
••2•0•0•1•1•0
••3•1•1•1•0•0
••4•1•0•0•0•1
••5•0•1•0•0•0
```

This is perhaps a little difficult to interpret at first sight, and you will have to remember that vectors are *ordered*. This output tells you how often which combinations of numbers in g and h occur. Beginning in the upper left corner: there is no occasion at which there is a 1 in g and a 1 in h. There is one occasion at which there is a 1 in g and a 2 in h (the first element of both vectors). There are no occasions where there is 1 in g and a 3 or a 5 or a 6 in h. Analogously, there are no positions where there's a 2 in g and a 1 or a 2 in h, but there is one position where g is 2 and h is 3 (the second position). And so on.

As you can probably imagine, `table` will prove very useful later on because it will be used to generate frequency lists as well as in the chapter on statistical evaluation. It is also worth mentioning here that there is a related function called prop.table, which provides a table of percentages. This function takes as its first argument a table generated by table and then as a second argument:

- a "1" if you want percentages that add up to 100 (or 1, for that matter) row-wise;
- a "2" if you want percentages that add up to 100 (or 1, for that matter) column-wise;
- nothing, if you want percentages that add up to 100 (or 1, for that matter) in the whole table.

Try it out by entering prop.table(table(g,•h))¶ or prop.table(table(g,•h),•1)¶ or prop.table(table(g,•h),•2)¶ ...

Let me finally mention two useful functions concerned with the order of elements in vectors. The first function is used to sort the elements of a vector in alphabetical or numerical order. It is appropriately called sort and the main two arguments we will discuss here are the vector to be sorted and an argument decreasing=F (the default setting) or decreasing=T:

```
>•h¶
[1]•2•3•1•5•2•6•3•1•2
>•sort(h,•decreasing=T)¶
[1]•6•5•3•3•2•2•2•1•1
```

The other function is called order. It takes as arguments one or more vectors and decreasing= . . . —but provides a very different output. Can you recognize what order does?

```
>•z<-c(3,•5,•10,•1,•6,•7,•8,•2,•4,•9)¶
>•order(z,•decreasing=F)¶
[1]••4••8••1••9••2••5••6••7•10••3
```

 **THINK**

BREAK

The output of order when applied to a vector z is a vector, which provides the order of the elements of z when they are sorted as specified. Let me clarify this rather opaque characterization by means of this example. If you want to sort the values of z in increasing order, you first have to take z's fourth value (which is the smallest value, 1). Thus, the first value of order(z,•decreasing=F)¶ is 4. The next value you have to take is the eighth value of z, which is 2. The next value you must take is the first value of z, namely 3, etc. The last value of z you take is its third one, 10, the maximum value of z. (If you provide order with more than one vector, additional vectors are used to break ties.) As you will see below, this function will turn out to be particularly handy when applied to data frames.

3.2.4 Saving Vectors

The most powerful function of outputting vectors is cat, which can also take on a variety of arguments, some of which we will discuss here:

```
cat("..."•,•file="",•sep="•",•append=F)
```

- The first, obligatory, argument "..." is the vector you want to output.
- The second argument is the `file` argument. If you don't provide it, the output is directed to the console. If you want to save the output into a file on Windows, you can either specify a path to a file directly or enter `choose.files()` as the argument. If you want to save the output into a file on MacOS or Linux, you can also either specify a path to a file directly or enter, say, `scan(nmax=1,•what="char")` so that you get prompted for a path. In general, make sure you provide an extension to the file name; in the context of this book, the files you create with `cat` or `write.table` below should have the extension ".txt".
- Then, the `sep` argument works in just the same way as it does in `scan`. If you want each vector element to get its own line, which is probably the most useful strategy for nearly all our cases, enter `sep="\n"`, but the default separator is just a space.
- The append argument specifies whether the output is appended to the end of an already existing file or whether a new file is created for the output (the default).

> You should now do "Exercise box 3.1: Handling vectors" ...

For further study/exploration:

- on how to test whether any one or all elements of a vector satisfy a logical expression: `?any¶` and `?all¶`
- the `fill`-argument of `cat`
- if at a more advanced stage, you write scripts whose execution takes a while, you can write this at the end of the script: `cat("\a")¶`. R will then produce the sound of a bell and you know the script is finished
- Spector (2008: Sections 6.1–6.4)

3.3 Factors

The second data structure we will look at are factors. Factors are superficially similar to vectors. They will only become important when we deal with reading and evaluating tables prepared in spreadsheet software. We will deal with them here only cursorily and return to them in Chapter 5. The most straightforward way to generate a factor is by first generating a vector as introduced above and then turning it into a factor using the command `factor` with the vector to be factorized as the first argument:

```
>•f<-c("open",•"open",•"open",•"closed",•"closed")¶
>•(f<-factor(f))¶
[1]•open•••open•••open•••closed•closed
Levels:•closed•open
```

```
>•is.factor(f)¶
[1]•TRUE
```

The function `factor` can also take a second argument `levels`, which specifies the levels the factor has. If you print out a factor, all levels of the factor that occur at least once are outputted just like `unique` outputs all values of a vector.

For further study/exploration: Spector (2008: Chapter 5)

3.4 Data Frames

While vectors are the most relevant data structure for retrieving data from corpora, data frames are the most relevant data structure for the evaluation of the data you annotated in a spreadsheet program. This data structure, which basically corresponds to a two-dimensional table, will be illustrated in this section.

3.4.1 Generating Data Frames

While data frames are usually loaded from text files generated with other programs, you can also generate data frames within R by combining several equally long vectors. Let us assume you have characterized five parts of speech in terms of three variables (or parameters or characteristics):

- a variable "TokenFrequency", i.e. the frequency of all individual words of this part of speech in a very small corpus C;
- a variable "TypeFrequency", i.e. the number of all different words of this part of speech in C;
- a variable "Class", i.e. whether the part of speech is an open class or a closed class.

Let us further assume the variables and the data frame you wish to generate should look like:

```
PartOfSpeech    →    TokenFrequency    →    TypeFrequency    →    Class¶
ADJ        →         421        →         271        →         open¶
ADV        →         337        →         103        →         open¶
N          →         1411       →         735        →         open¶
CONJ       →         458        →         18         →         closed¶
PREP       →         455        →         37         →         closed¶
```

Figure 3.5 An example data frame.

In a first step, you generate the four vectors, one for each column of the data frame:

```
>•rm(list=ls(all=T))¶
>•PartOfSpeech<-c("ADJ",•"ADV",•"N",•"CONJ",•"PREP")¶
>•TokenFrequency<-c(421,•337,•1411,•458,•455)¶
>•TypeFrequency<-c(271,•103,•735,•18,•37)¶
>•Class<-c("open",•"open",•"open",•"closed",•"closed")¶
```

Note that the first row of the data frame you want to generate does not contain data points, but the names of the columns. You must now also decide whether you would like the first column to contain data points or the names of the rows. If you prefer the former, then you can simply generate the data frame using data.frame, which in this case takes as arguments only the vectors/factors you want to combine:

```
>•x<-data.frame(PartOfSpeech,•TokenFrequency,•TypeFrequency,•Class)¶
```

Note that the order of the vectors is only important in that it determines the order of the columns in the data frame.

Let us now look at the data frame and its properties:

```
>•x¶
••PartOfSpeech•TokenFrequency•TypeFrequency••Class
1••••••••••ADJ••••••••••••421••••••••••••271•••open
2•••••••••ADV••••••••••••337••••••••••••103•••open
3••••••••••N••••••••••••1411••••••••••••735•••open
4•••••••••CONJ••••••••••••458••••••••••••18•closed
5•••••••••PREP••••••••••••455••••••••••••37•closed

>•str(x)¶
'data.frame':•••5•obs.•of••4•variables:
•$•PartOfSpeech••:•Factor•w/•5•levels•"ADJ","ADV","CONJ",..:•1•2•4•3•5
•$•TokenFrequency:•num•••421••337•1411••458••455
•$•TypeFrequency•:•num••271•103•735•18•37
•$•Class•••••••••:•Factor•w/•2•levels•"closed","open":•2•2•2•1•1
```

We can see several things from this output. First, R has generated the data frame as desired. Second, R has automatically converted those vectors that contained character strings into factors (viz. PartOfSpeech and Class), and we can see that factors are internally represented as numbers (e.g. for Class, "closed" is 1 and "open" is 2). Third, since we have not specified any row names, R automatically numbers the rows. Finally, the column names in the output of str are preceded by a $ sign. This means that when you have stored a data frame in R, you can access columns by using the name of the data frame, a $ sign, and a column name:

```
>•x$PartOfSpeech¶
[1]•ADJ••ADV••N••••CONJ•PREP
Levels:•ADJ•ADV•CONJ•N•PREP
```

The above way of generating data frames is the default you will use in most cases. If, for some reason, you prefer to use, say, the parts of speech as the row names, you have to slightly change the way you invoke data.frame:

```
>•(x.2<-data.frame(TokenFrequency,•TypeFrequency,•Class,•
  row.names=PartOfSpeech))¶
•••••TokenFrequency•TypeFrequency••Class
ADJ•••••••••••••421•••••••••••271•••open
ADV•••••••••••••337•••••••••••103•••open
N•••••••••••••1411•••••••••••735•••open
CONJ•••••••••••458•••••••••••18•closed
PREP•••••••••••455•••••••••••37•closed

>•str(x.2)¶
'data.frame':•••5•obs.•of••3•variables:
•$•TokenFrequency:•num•••421••337•1411••458••455
•$•TypeFrequency••:•num••271•103•735•18•37
•$•Class•••••••••:•Factor•w/•2•levels•"closed","open":•2•2•2•1•1
```

Now the data frame contains just three variables, since the parts of speech constitute the row names. Note that this is only possible if all elements of the vector singled out for the row names are different from each other: R does not allow duplicate row names.

3.4.2 Loading and Saving Data Frames

The more common way of getting R to recognize a data frame is to load a file that was generated with spreadsheet software. Let us assume you have generated a spreadsheet file <C:/_qclwr/_input files/dat_dataframe-a.ods> containing the above table x (using a text editor or a spreadsheet software such as OpenOffice.org Calc). Ideally your spreadsheet would only contain alphanumeric characters and no whitespace characters (to facilitate later handling in R or other software). The first step is to save that file as a raw text file. In OpenOffice.org 3.0 Calc, you choose the menu "*File: Save As*" and choose "Text CSV (.csv)" from the *Save as type* menu, and deactivate the automatic file extension. Then, you enter a file name *with* the extension ".txt" and confirm you want to save the data into a .csv file, if prompted to do so. After that, choose "{Tab}" as the field delimiter and, usually, delete the text delimiter.

The second step is to load this file <C:/_qclwr/_inputfiles/dat_dataframe-a.txt> into R with read.table. These are the most frequently used arguments of read.table with their default settings:

```
read.table(file=…,•header=F,•sep="",•quote="\"‘",•dec=".",•row.names,•
  na.strings="NA",•comment.char="#")
```

- The argument file is obligatory and specifies a data frame as saved above (as before with vectors, you can also just enter choose.files() or file.choose() or scan-(nmax=1,•what="char") to be prompted to choose a file interactively.
- The argument header specifies whether the first row contains the labels for the columns (header=T)—as would normally be the case—or not (header=F).
- The argument sep, as mentioned above, specifies the character that separates data fields, and given the file we wish to import—and in fact most of the time—our setting should be sep="\t" for a tab.
- We also know the dec argument, which is used as introduced above; in an English locale, the setting would not have to be changed.
- The argument quote, which provides the characters used for quoted strings. On most if not all occasions, you should set this to quote="" to avoid input problems.
- The argument row.names can either be a vector containing names for the rows or, more typically, the number of the columns containing the row names (typically 1). If you do not specify a row.names argument, the rows will be numbered automatically.
- The argument na.strings takes a character vector of strings that are to be considered as unavailable/missing data.[8]
- Finally, the argument comment.char, which provides R with the character that separates comments from the rest of the line. Just like with quote, you will most likely wish to set this to comment.char="".

Thus, in order to import the table we just saved from OpenOffice.org 3.0 Calc into a data frame called x, we enter the following at the R console:

```
>•rm(list=ls(all=T))¶
>•x<-read.table(choose.files(),•header=T,•sep="\t",•comment.char="")•#•no•
  row.names:•R•numbers•rows,•or¶
>•x.2<-read.table(choose.files(),•header=T,•row.names=1,•sep="\t",•
  comment.char="")•#•with•row.names¶
```

If, by contrast, you wish to save a data frame from R into a text file, you need write.table. Here are the most important arguments and their default settings:

```
write.table(x,•file=. . .,•append=F,•sep="•",•eol="\n",•na="NA",•dec=".",•
  quote=T,•row.names=T,•col.names=T)
```

- The argument x is the data frame to be saved.
- The arguments file, append, sep, and dec are used in the ways introduced above.
- The argument quote specifies whether you want factor levels within double quotes, which is usually not particularly useful for editing data in a spreadsheet software.

- The argument eol provides the character that marks the end of lines. This will normally be eol="\n", which is also the default.
- The arguments row.names and col.names specify whether you would like to include the names (or numbers) of the rows and the names of the columns in the file.

Given the above default settings and under the assumption that your operating system uses an English locale, there are two most common ways to save such data frames: if you have a data frame without row names such as x, you enter the following line:

```
>•write.table(x,•file="C:/_qclwr/_outputfiles/03-4-2_df-a.txt"),•quote=F,•
  sep="\t",•row.names=F)¶
```

Note how row.names=F suppresses R's row numbering. If you have a data frame with row names such as x.2, enter the same but add the argument col.names=NA, which makes sure that the column names stay in place:

```
>•write.table(x.2,•file="C:/_qclwr/_outputfiles/03-4-2_df-b.txt"),•quote=F,•
  sep="\t",•col.names=NA)¶
```

The default setting for the decimal separator can be changed in OpenOffice.org Calc (*Tools: Options: Language Settings: Languages*), in Microsoft Excel (*Tools: Options: International*), or—for the locale—in Windows (Control panel: Regional and language options).

For further study/exploration:

- on how to handle and detect missing or non-numeric data: ?NA¶ and ?is.na¶ as well as ?NaN¶ and ?is.nan¶, ?na.action¶, ?na.omit¶, and ?na.fail¶
- on how to test whether cases (e.g. rows in data frames) are complete: ?complete.cases¶

3.4.3 Accessing and Processing (Parts of) Data Frames

There are several R functions that are useful for accessing parts of data frames that are to be used for subsequent analysis. To look at these, first close R, then start a new session and load the data frame from <C:/_qclwr/_inputfiles/dat_dataframe-a.txt> above into x again:

```
>•x<-read.table(choose.files(),•header=T,•sep="\t",•;comment.char="")¶
```

One easy way to access columns in a data frame involves the dollar sign (cf. above).

```
>•str(x)¶
'data.frame':•••5•obs.•of••4•variables:
•$•PartOfSpeech••:•Factor•w/•5•levels•"ADJ","ADV","CONJ",..:•1•2•4•3•5
•$•TokenFrequency:•num•••421••337•1411••458••455
•$•TypeFrequency•:•num••271•103•735•18•37
•$•Class•••••••••:•Factor•w/•2•levels•"closed","open":•2•2•2•1•1
>•x$TokenFrequency¶
[1]••421••337•1411••458••455
>•x$Class¶
[1]•open•••open•••open•••closed•closed
Levels:•closed•open
```

A simpler possibility to achieve the same objective is attach.

```
>•attach(x)¶
```

R does not generate output, but as a result of the function you can now access the variables as defined by the columns. (To undo an attach(. . .)¶, use detach(. . .)¶.)

```
>•Class¶
[1]•"open"•••"open"•••"open"•••"closed"•"closed"
Levels:•closed•open
```

It is crucial to note that these variables are, so to speak, copies of the original ones in the data frame. That means you can change them,

```
>•(TokenFrequency[4]<-20)¶
[1]••421••337•1411•••20••455
```

. . . but note that these changes do not affect the data frame they come from:

```
>•x¶
••PartOfSpeech•TokenFrequency•TypeFrequency••Class
1•••••••••••ADJ••••••••••••421••••••••••••271•••open
2•••••••••••ADV••••••••••••337••••••••••••103•••open
3•••••••••••N••••••••••••1411••••••••••••735•••open
4•••••••••CONJ••••••••••••458•••••••••••••18•closed
5•••••••••PREP••••••••••••455•••••••••••••37•closed
```

In order for that to happen you would have to write x$TokenFrequency[4]<-20¶.[9] But let us take this change of TokenFrequency back now:

```
>•TokenFrequency[4]<-458¶
```

The most versatile approach is again subsetting. We saw above that we can use square brackets to select parts of unidimensional vectors. Since data frames are two-dimensional data structures, we now need two (sets of) figures, one for the rows and one for the columns:

```
>•x[2,3]•#•the•value•of•the•second•row•and•the•third•column¶
[1]•103
>•x[2,]•#•all•values•of•the•second•row•(because•no•column•is•specified)¶
••PartOfSpeech•TokenFrequency•TypeFrequency••Class
2•••••••••ADV••••••••••••337••••••••••••103•••open
>•x[,3]•#•all•values•of•the•third•column•(because•no•row•is•specified)¶
[1]•271•103•735••18••37
>•x[2:3,4]•#•two•values•of•the•fourth•column¶
[1]•open•open
Levels:••closed•open
>•x[c(1,3),•c(2,4)]•#•the•1st•and•3rd•row•of•the•2nd•and•4th•column¶
••TokenFrequency•Class
1•••••••••••••421••open
3•••••••••••1411••open
```

As you can see, row and column names are not counted. Also, recall the versatile ways of juxtaposing different functions:

```
>•which(x[,2]>450)¶
[1]•3•4•5
>•x[,3][which(x[,3]>100)]¶
[1]•271•103•735
>•x[,3][x[,3]>100]¶
[1]•271•103•735
>•TypeFrequency[TypeFrequency>100]¶
[1]•271•103•735
```

Sometimes, you want to investigate only one part of a data frame, namely a part where one variable has a particular value. One way to get at a subset of the data follows logically from what we have done already. Let us assume you want to define a data frame y that contains only those rows of x referring to open-class words. If you have already made the variables of a data frame available using attach, this would be one possibility to do that:

```
>•(y<-x[which(Class=="open"),])•#•or•shorter:•
  (y<-x[Class=="open",])¶
••PartOfSpeech•TokenFrequency•TypeFrequency•Class
```

```
1·········ADJ············421············271··open
2·········ADV············337············103··open
3··········N············1411············735··open
```

If you have not made the variable names available, you can still do this:

```
>·(y<-x[which(x[,4]=="open"),])·#·or·shorter:·
  (y<-x[x[,4]=="open",])¶
··PartOfSpeech·TokenFrequency·TypeFrequency·Class
1·········ADJ············421············271··open
2·········ADV············337············103··open
3··········N············1411············735··open
```

A doubtlessly more elegant way of achieving the same is provided by subset. This function takes as its first argument the data frame to subset and as additional arguments a logical expression specifying the selection conditions.

```
>·(y<-subset(x,·Class=="open"))¶
··PartOfSpeech·TokenFrequency·TypeFrequency·Class
1·········ADJ············421············271··open
2·········ADV············337············103··open
3··········N············1411············735··open

>·(y<-subset(x,·Class=="open"·&·TokenFrequency<1000))¶
··PartOfSpeech·TokenFrequency·TypeFrequency·Class
1·········ADJ············421············271··open
2·········ADV············337············103··open

>·(y<-subset(x,·PartOfSpeech·%in%·c("ADJ",·"ADV")))¶
··PartOfSpeech·TokenFrequency·TypeFrequency·Class
1·········ADJ············421············271··open
2·········ADV············337············103··open
```

This way data frames can be conveniently customized for many different ways of analysis. If, for whatever reason, you wish to edit data frames in R rather than in a spreadsheet program—for example, because Microsoft Excel 2003 and OpenOffice.org 3.0 Calc are limited to data sets with maximally 65,536 rows, a ridiculously small number for many corpus-linguistic purposes—you can also do this in R.[10] The function fix takes as an argument the name of a data frame and opens a spreadsheet-like interface in which you can edit individual entries. After the editing operations have been completed, you can then save the data frame for further processing.

Finally, let me mention a very practical way of reordering data frames by means of order introduced above. Recall that order returns a vector of positions of elements and recall that subsetting can be used to access parts of vectors, data frames, and lists. As a matter of fact, you may already guess how you can reorder data frames. For example, imagine you want to sort the rows in the data frame x according to Class (ascending in alphabetical order) and, within Class, according to values of the column TokenFrequency (in descending order). Of course we can use order to achieve this goal, but there is one tricky issue involved here: the default ordering style of order is decreasing=F, i.e. in ascending order, but you actually want to apply two different styles: ascending for Class and descending for TokenFrequency, so changing the default alone will not help. What you can do, though, is this:

```
>•ordering.index<-order(Class,•-TokenFrequency);•ordering.index¶
[1]•4•5•3•1•2
```

That is, you apply order not to TokenFrequency, but to the negative values of TokenFrequency, effectively generating the desired descending sorting style. Then, you can use this vector to reorder the rows of the data frame by subsetting:

```
>•x[ordering.index,]¶
••PartOfSpeech•TokenFrequency•TypeFrequency••Class
4•••••••••CONJ•••••••••••••458•••••••••••18•closed
5•••••••••PREP•••••••••••••455•••••••••••37•closed
3•••••••••••N••••••••••••1411•••••••••••735•••open
1•••••••••ADJ•••••••••••••421•••••••••••271•••open
2•••••••••ADV•••••••••••••337•••••••••••103•••open
```

Of course, this could have been done in one line:[11]

```
>•x[order(Class,•-TokenFrequency),]¶
```

You can also apply a function we got to know above, sample, if you want to reorder a data frame randomly. This may be useful, for example, to randomize the order of stimulus presentation in an experiment or to be able to randomly select only a part of the data in a data frame for analysis. One way of doing this is to first retrieve the number of rows that need to be reordered using dim, whose first value will be the number of rows and the second the number of columns (just as in subsetting or prop.table):

```
>•no.of.rows<-dim(x)[1]•#•for•the•columns:•
  no.of.columns<-dim(x)[2]¶
```

Then, you reorder the row numbers randomly using sample and then you use subsetting to change the order; of course you may have a different (random) order:

```
>•(ordering.index<-sample(no.of.rows))¶
[1]•1•4•2•3•5
>•x[ordering.index,]¶
••PartOfSpeech•TokenFrequency•TypeFrequency••Class
1•••••••••ADJ•••••••••••••421•••••••••••271•••open
```

```
4·········CONJ············458············18·closed
2··········ADV············337············103···open
3············N············1411············735···open
5·········PREP············455············37·closed
```

Alternatively, you do it all in one line:

```
>·x[sample(dim(x)[1]),]¶
```

Finally, on some occasions you may want to sort a data frame according to several columns and several sorts (increasing and decreasing). However, you cannot apply a minus sign to *factors* to force a particular sorting style, which is why in such cases you need to use rank, which rank-orders the column factors first into numbers, to which then the minus sign can apply:

```
>·ordering.index<-order(-rank(Class),·-rank(PartOfSpeech))¶
>·x[ordering.index,]¶
··PartOfSpeech·TokenFrequency·TypeFrequency··Class
3············N············1411············735···open
2··········ADV············337············103···open
1··········ADJ············421············271···open
5·········PREP············455············37·closed
4·········CONJ············458············18·closed
```

> You should now do "Exercise box 3.2: Handling data frames" . . .

For further study/exploration:

- on how to edit data frames in a spreadsheet interface within R: ?fix¶
- on how to merge different data frames: ?merge¶
- on how to merge different columns and rows into matrices (which can in turn be changed into data frames easily): ?cbind¶ and ?rbind¶
- on how to access columns of data frames without using attach: ?with¶
- Spector (2008: Chapter 2, Section 6.8)

3.5 Lists

While the data frame is probably the most central data structure for statistical evaluation in R and has thus received much attention here, data frames are actually just a special kind of data structure, namely lists. More specifically, data frames are lists that contain vectors and factors, which all have the same length. Lists are a much more versatile data structure, which can in turn contain various different data structures within them. For example, a list can contain different kinds of vectors, data frames, other lists, as well as data structures that are not discussed here:

```
>•rm(list=ls(all=T))¶
>•a.vector<-c(1:10)•#•generates•a•vector•with•the•;numbers•from•one•to•ten¶
>•a.dataframe<-read.table(choose.files(),•header=T,•sep="\t",•
  comment.char="")•#•load•the•dataframe•we•used•in•Section•3.4¶
>•another.vector<-c("This",•"may",•"be",•"a",•"sentence",•"from",•"a",•
  "corpus",•"file",•".")¶
>•(a.list<-list(a.vector,•a.dataframe,•another.vector))¶
[[1]]
•[1]••1••2••3••4••5••6••7••8••9•10
[[2]]
••PartOfSpeech•TokenFrequency•TypeFrequency••Class
1••••••••••ADJ•••••••••••••421•••••••••••271•••open
2••••••••••ADV•••••••••••••337•••••••••••103•••open
3•••••••••••N••••••••••••1411•••••••••••735•••open
4••••••••CONJ•••••••••••••458•••••••••••18•closed
5••••••••PREP•••••••••••••455•••••••••••37•closed
•[[3]]
•[1]•"This"•••••"may"•••••••"be"•••••••"a"•••••••"sentence"•"from"
•[7]•"a"•••••••••"corpus"•••"file"•••••"."
```

As you can see, the three different elements—a vector of numbers, a data frame with different columns, and a vector of character strings—are now stored in one data structure:

```
>•str(a.list)¶
List•of•3
•$•:•int•[1:10]•1•2•3•4•5•6•7•8•9•10
•$•:'data.frame':•••••••5•obs.•of••4•variables:
••..$•PartOfSpeech••:•Factor•w/•5•levels•"ADJ","ADV","CONJ",..:•1•2•4•3•5
••..$•TokenFrequency:•int•[1:5]•421•337•1411•458•455
••..$•TypeFrequency•:•int•[1:5]•271•103•735•18•37
••..$•Class•••••••••:•Factor•w/•2•levels•"closed","open":•2•2•2•1•1
•$•:•chr•[1:10]•"This"•"may"•"be"•"a"•...
```

An alternative way to generate such a list would consist of labeling the elements of the list, i.e. giving them names just like above when we gave names to elements of vectors. You can do that either when you generate the list . . .:

```
>•a.list<-list(Part1=a.vector,•Part2=a.dataframe,•Part3=another.vector)¶
```

. . . or you can do the labeling with names as you did with vectors above:

```
>•names(a.list)<-c("Part1",•"Part2",•"Part3")¶
```

But let's for now stick to the unlabeled format (and thus redefine a.list as before):

```
>•a.list<-list(a.vector,•a.dataframe,•another.vector)•#•redefine•a.list¶
```

This kind of data structure is interesting because it can contain many different things and because you can access individual parts of it in ways similar to those you use for other data structures. As the above output indicates, the elements in lists are numbered and the numbers are put between double square brackets. Thus:

```
>•a.list[[1]]¶
[1]••1••2••3••4••5••6••7••8••9•10
>•a.list[[2]]¶
••PartOfSpeech•TokenFrequency•TypeFrequency••Class
1•••••••••ADJ•••••••••••421•••••••••••271•••open
2•••••••••ADV•••••••••••337•••••••••••103•••open
3•••••••••••N•••••••••••1411•••••••••••735•••open
4•••••••CONJ•••••••••••458•••••••••••18•closed
5•••••••PREP•••••••••••455•••••••••••37•closed
>•a.list[[3]]¶
[1]•"This"•••••"may"••••••"be"•••••••"a"••••••••"sentence"
[6]•"from"•••••"a"•••••••••"corpus"•••"file"•••••"."
```

It is important to bear in mind that there are two ways of accessing a list's elements. The one with double square brackets is the one suggested by the output of a list. This way, you get each element as the kind of data structure you entered into the list:

```
>•is.list(a.list[[1]])¶
[1]•FALSE
>•is.vector(a.list[[1]])¶
[1]•TRUE
>•is.data.frame(a.list[[2]])¶
[1]•TRUE
>•is.vector(a.list[[3]])¶
[1]•TRUE
```

However, there is also the general subsetting approach, which you know uses single square brackets. Can you see the difference in the notation with double square brackets? If you look at the two data structures—the one you get with single square brackets and the one you get with double square brackets—what is happening here?

```
>•a.list[1]¶
[[1]]
[1]••1••2••3••4••5••6••7••8••9•10
>•is.list(a.list[1])¶
[1]•TRUE
```

THINK

BREAK

The difference is this: if you access a component of a list with *double* square brackets, you get that component as an instance of the kind of data structure that was entered into the list (cf. the first output). If you access a component of a list with *single* square brackets, you get that component as an instance of the data structure from which you access it, i.e. as a list. This may seem surprising or confusing, but it is actually only consistent with what you already know: if you use single square-bracket subsetting on a vector, you get a vector, if you use single square-bracket subsetting (with more than one row and column chosen) on a data frame, you get a data frame, etc. So, once you think about it, no big surprise.

```
>•a.list[[1]]¶
[1]••1••2••3••4••5••6••7••8••9•10
>•a.list[1]¶                           unlike above, this tells you
[[1]]  ←————————————————————           you're looking at a list
[1]••1••2••3••4••5••6••7••8••9•10
```

Finally, if you named the elements of a list, you could also use those (but we didn't, which is why you get NULL):

```
>•a.list$Part1•#•or•a.list[["Part1"]]¶
NULL
```

Of course, you can also access several elements of a list at the same time. Since you will get a list as output, you use the by now familiar strategy of single square brackets:

```
>•a.list[c(1,3)]¶
[[1]]
[1]••1••2••3••4••5••6••7••8••9•10
[[2]]
[1]•"This"•••••"may"••••••"be"•••••••"a"••••••••"sentence"
[6]•"from"•••••"a"•"corpus"•••"file"•••••"."
```

So far we looked at how you access *elements* of lists, but how do you access *parts of elements* of lists? The answer of course uses the notation with double square brackets (because you now do not want a list, but, say, a vector) . . .

```
>•a.list[[c(1,3)]]•#•take•the•third•part•of•the•first•element•of•the•list¶
[1]•3
```

. . . which is the same as the probably more intuitive way . . .

```
>•a.list[[1]][3]•#•take•the•third•part•of•the•first•element•of•the•list¶
[1]•3
```

. . . and this is how you access more than part of a list at the same time:

```
>•a.list[[1]][3:5]•#•consecutive•parts¶
[1]•3•4•5
>•a.list[[1]][c(3,•5)]•#•note•that•"a.list[[1]][3,•5]"•does•not•work•
   because•this•would•try•to•access•row•3,•column•5•of•a.list[[1]]¶
[1]•3•5
```

Thus, this is how to handle data frames in lists such as, for instance, accessing the second element of the list, and then the value of its third row and second column:

```
>•a.list[[2]][3,2]¶
[1]•1411
>•a.list[[2]][3,2:4]¶
••TokenFrequency•TypeFrequency•Class
3•••••••••••1411•••••••••••735••open
```

And this is how you take the second element of the list, and then the elements in the third row and the second and fourth column:

```
>•a.list[[2]][3,c(2,•4)]¶
••TokenFrequency•Class
3•••••••••••1411••open
```

As can now be seen, once one has figured out which bracketing to use, lists can be an extremely powerful data structure. They can be used for very many different things. One is that many statistical functions output their results in the form of a list so if you want to be able to access your results in the most economical way, you need to know how lists work.

A second one is that sometimes lists facilitate the handling of your data. For example, we have used the data frame a.dataframe in this section, and above we introduced subset to access parts of a data frame that share a set of variable values or levels. However, for larger or more complex analyses, it may also become useful to be able to split up the data frame into smaller parts depending on the values some variable takes. You can do this with split, which takes as its first argument the data frame to be split up and as its second argument the name of the variable (i.e. column name) according to which the data frame is to be split. Why do I mention this here? Because the result of split is a list:

```
>•x<-a.list[[2]]¶
>•(y<-split(x,•Class))¶
$closed
••PartOfSpeech•TokenFrequency•TypeFrequency••Class
4•••••••••CONJ•••••••••••••458••••••••••••18•closed
5•••••••••PREP•••••••••••••455••••••••••••37•closed
$open
••PartOfSpeech•TokenFrequency•TypeFrequency•Class
1••••••••••ADJ•••••••••••••421•••••••••••271••open
2••••••••••ADV•••••••••••••337•••••••••••103••open
3•••••••••••N••••••••••••1411•••••••••••735••open
```

This then even allows you to access the parts of the list using the above $ notation:

```
>•y$open¶
••PartOfSpeech•TokenFrequency•TypeFrequency•Class
1••••••••••ADJ•••••••••••••421•••••••••••271••open
2••••••••••ADV•••••••••••••337•••••••••••103••open
3•••••••••••N••••••••••••1411•••••••••••735••open
```

Note that you can also split up a data frame according to all combinations of two variables (i.e. more than one variable), the second argument of split must be a list of these variables, which of course does not make too much sense in the present example:

```
>•split(x,•list(Class,•PartOfSpeech))¶
```

One final piece of advice: since lists are such a versatile data structure, you may often consider storing all your data, all your code, and all your results in one list and save that as a file to which you can always return and from which you can always retrieve all of your material easily.

3.6 Elementary Programming Functions

In order to really appreciate the advantages R has to offer over nearly all available corpus-linguistic software, we will now introduce a few immensely important functions, control statements, and control-flow structures that will allow you to apply individual functions as well as sets of functions to data more often than just the single time we have so far been covering. To that end, this section will first introduce conditional expressions; then we will look at the notion of loops; finally, we will turn very briefly to a few functions which can sometimes replace loops and are often more useful because they are faster and less memory consuming.

3.6.1 Conditional Expressions

You will often be in the situation that you want to execute a particular (set of) function(s) if and only if some condition *C* is met, and that you would like to execute some other (set of) function(s) if *C* is not met. The most fundamental way of handling conditions in R involves a control-flow structure that can be expressed as follows. Recall from above the fact that the plus sign at the beginning of lines is not something you enter but the version of the prompt that R displays when it expects further input before executing a (series of) command(s):

```
if•(some•condition)•{¶
•••execute•some•function(s)•F¶
}•else•if•(other•condition)•{¶
•••execute•some•other•function(s)•G¶
}•else•{¶
•••execute•yet•some•other•function(s)•H¶
}¶
```

(Note that the structures else•if•{•...•} and else•{•...•} are optional.) The expression condition represents something we have already encountered above (e.g. in the context of which but also elsewhere), namely an expression with logical operators that can be evaluated and that either returns TRUE or FALSE. If it returns TRUE, R will execute the first (set of) function(s), the ones labeled *F* above. If the expression condition returns FALSE, R will test whether the second condition is true. If so, it will execute the (set of) function(s) labeled *G* above, otherwise it will execute the (set of) function(s) labeled *H*. Two simple examples will clarify this:

```
>•a<-2•#•you•can•of•course•insert•another•number•here¶
>•if•(a>2)•{¶
+••••cat("a•is•greater•than•2.\n")¶
+•}•else•{¶
+••••cat("a•is•not•greater•than•2.\n")¶
+•}¶
a•is•not•greater•than•2.
```

and:

```
>•b<-"car"•#•you•can•of•course•insert•another•word•here¶
>•if•(b=="automobile")•{¶
+••••cat("b•is•'automobile'!\n")¶
+•}•else•if•(b=="car")•{¶
+••••cat("b•is•'car'!\n")¶
+•}•else•{¶
+••••cat("b•is•neither•'automobile'•nor•'car'.\n")¶
+•}¶
b•is•'car'!
```

You may wonder what the series of leading spaces in some lines are for, especially since I said above that R only cares about space within quotes. This is still correct: R does not care about these spaces, but *you* will. The spaces' only function is to enhance the legibility and interpretability of your small script. In the remainder of the book, every indentation by three spaces represents one further level of embedding within the script. The third line of the second example above is indented by three spaces, which shows you that it is a line within one conditional expression (or loop, as we will see shortly). When you write your own scripts, this convention will make it easier for you to recognize coherent blocks of code which all belong to one conditional expression. Of course, you could also use two spaces, a tab stop, or . . .

When do you need conditional expressions? Well, while not all of the following examples are actually best treated with if, they clearly indicate a need for being able to perform a particular course of action only when particular conditions are met. You may want to:

- include a corpus file in your analysis if the corpus header reveals it is a file containing spoken language;
- search a line of a corpus file if it belongs to a particular utterance;
- include a word in your frequency list if its frequency reaches a particular value, etc.

You should now do "Exercise box 3.3: Conditional expressions" . . .

For further study/exploration:

- on how to test an expression and specify what to do when the expression is true or false in one line: ?ifelse¶
- on how to choose one of several alternatives and perform a corresponding action: ?switch¶

3.6.2 Loops

This section introduces some basic loops. Loops are one useful way to execute one or more functions several times. While R offers several different ways to use loops, I will only introduce one of them here in a bit more detail, namely for-loops, and show just two very small examples for while and repeat-loops.

A for-loop has the following structure:

```
for•(var•in•seq)•{¶
•••execute•some•function(s)¶
}¶
```

This requires some explanation. The expression var stands for any variable's name, and seq stands for anything that can be interpreted as a sequence of values (where one value is actually enough to constitute a sequence). Let us look at the probably simplest conceivable example, that of printing out numbers:

```
>•for•(i•in•1:3)•{¶
+••••cat(i,•"\n")¶
+•}¶
1
2
3
```

This can be translated as follows: generate a variable called i as an index for looping. The value i should take on at the beginning is 1, and then R executes the function(s) within the loop—i.e. those enclosed in curly brackets. Here, this just means printing i and a newline. When the closing curly bracket is reached, i should take on the next value in the user-defined sequence and again perform the functions within the loop, etc. until the functions within the loop have been executed with i having the last value of the sequence. Then, the for-loop is completed/exited. The above kind of sequence is certainly the most frequently used one—the index i starts with 1 and proceeds through a series of consecutive integers—but it is by no means the only one:

```
>•j<-seq(2,•1,•-0.5)¶
>•for•(i•in•j)•{¶
+••••cat(i,•"\n")¶
+•}¶
2
1.5
1
```

Here, i takes on all values of j. Of course, you can also use nested loops, that is you can execute one loop within another one, but be aware of the fact that you must use different names for the looping indices (i and j in this example):

```
>•for•(i•in•1:2)•{¶
+••••for•(j•in•6:7)•{¶
+••••••••cat(i,•"times",•j,•"is",•i*j,•"\n")¶
+••••}¶
```

```
+•}¶
1•times•6•is•6
1•times•7•is•7
2•times•6•is•12
2•times•7•is•14
```

While these examples only illustrate the most basic kinds of loops conceivable (in the sense of what is done within each iteration), we will later use for-loops in much more complex and creative ways.

Since you specify a sequence of iterations, for-loops are most useful when you must perform a particular operation a known number of times. However, sometimes you do not know the number of times an operation has to be repeated. For example, sometimes the number of times something has to be done may depend on a particular criterion. R offers several easy ways of handling these situations. One possibility of, so to speak, making loops more flexible is by means of the control expression next. If you use next within a loop, R skips the remaining instructions in the loop and advances to the next element in the sequence/counter of the loop. The following is a (rather boring) example:

```
>•for•(i•in•1:3)•{¶
+••••if•(i==2)•{¶
+••••••••next¶
+••••}¶
+••••cat(i,•"\n")¶
+•}¶
1
3
```

As you can see, when the logical expression is true, R never reaches the line in which i is printed. Another possibility is break. If you use break within a loop, the iteration is interrupted and R proceeds with the first statement *outside* of the innermost loop in which break was processed. This is how it works:

```
>•for•(i•in•1:5)•{¶
+••••if•(i==3)•{¶
+••••••••break¶
+••••}¶
+••••cat(i,•"\n")¶
+•}¶
1
2
```

This time, when the logical expression is true, R processes break and breaks out of the for-loop. Another way of approaching the same issue uses a loop structure other than a for-loop. For example, you could also use a so-called repeat-loop. A repeat-loop repeats a set of functions until you break out of it, that is, it does not have a pre-defined end-point:

```
>•i<-1¶
>•repeat•{¶
+••••cat(i,•"\n")¶
+••••i<-i+1•#•this•way•of•incrementing•a•variable•is•extremely•useful!¶
+••••if•(i==3)•{•break•}¶
+•}¶
1
2
```

A final way involves a control-flow construct called while, which executes a set of functions as long as a logical condition is true. For example, we can rewrite our last two examples using break in a while-loop, effectively saying "as long as i is smaller than 3, print it":

```
>•i<-1¶
>•while•(i<3)•{¶
+••••cat(i,•"\n")¶
+••••i<-i+1¶
+•}¶
1
2
```

Most of the time, you can use any of these functions, the differences only being cosmetic in nature. However, when you neither know how often you will have to execute a set of functions, nor know at least the maximum possible number of iterations or a break criterion, then you cannot use a for-loop.

Before we proceed, let me mention that loops slightly complicate your replicating scripts directly at the console. This is because once you enter a loop, R will not begin with the execution before the final "}" and you may not be able to recapitulate the inner working of the iterations. My recommendation to handle this is the following: when a loop begins like this, for•(i•in•1:3)•{¶, but you want to proceed through the loop stepwise (for instance, in order to find an error), then do *not* enter this line into R immediately. Rather, to see what is happening inside the loop, set the counter variable, i in this case, to its first value *manually* by writing i<-1¶. This way, i has been defined and whatever in the loop depends on i can proceed. If you then want to go through the loop once more, but do not yet want to do all iterations, just increment i by 1 manually (i<-i+1¶) and "iterate" again. Once you have understood the instructions in the loop and you want to execute it all in one go, just type the real beginning of the loop (for•(i•in•1:3)•{¶) or copy and paste from the script file. This way you will be able to understand more complex scripts more easily.

> **You should now do "Exercise box 3.4: Loops"** . . .

3.6.3 Rules of Programming

This book cannot serve as a fully-fledged introduction to good programming style (in R or in general). In fact, I have not even always shown the simplest or most elegant way of achieving a particular goal because I wanted to avoid making matters more complicated than absolutely necessary. However, given that there are usually many different ways to achieve a particular goal, there are two aspects of reasonable R programming that I would like to emphasize because they will help to:

- increase the speed with which some of the more comprehensive tasks can be completed;
- reduce memory requirements during processing; and
- reduce the risk of losing data when you perform more comprehensive tasks.

The first of the two aspects is concerned with capitalizing on the fact that R's most fundamental data structure is the vector, and that R can apply functions to all elements of a vector at once without having to explicitly loop over all elements of a vector individually. Thus, while conditional expressions and for-loops are extremely powerful control structures and you can of course force R to do many things using these structures, there are sometimes more elegant ways to achieve the same objectives. Why may some ways be more elegant? First, they may be more elegant because they achieve the same objective with (much) fewer lines of code. Second, they may be more elegant because they are more compatible with how the internal code of R has been optimized in terms of processing speed and allocation of memory. Let us look at a few examples using the vectors from above again:

```
>•PartOfSpeech<-c("ADJ",•"ADV",•"N",•"CONJ",•"PREP")¶
>•TokenFrequency<-c(421,•337,•1411,•458,•455)¶
>•TypeFrequency<-c(271,•103,•735,•18,•37)¶
>•Class<-c("open",•"open",•"open",•"closed",•"closed")¶
```

Let us assume we would like to determine the overall token frequencies of open class items and closed class items in our data set. There are several ways of doing this. One is terribly clumsy and involves using a for-loop to determine for each element of Class whether it is "open" or "closed":[12]

```
>•sum.closed<-0;•sum.open<-0•#•define•two•vectors•for•the•results¶
>•for•(i•in•1:5)•{¶
+••••current.class<-Class[i]•#•access•each•word•class¶
+••••if•(current.class=="closed")•{¶
+••••••••sum.closed<-sum.closed+TokenFrequency[i]•#•if•the•current•class•is•
   "closed",•add•its•token•frequency•to•the•first•result•vector¶
+••••}•else•{¶
+••••••••sum.open<-sum.open+TokenFrequency[i]•#•if•the•current•class•is•
   not•"closed",•add•its•token•frequency•to•the•other•result•vector¶
+••••}•#•end•of•if:•test•current•class¶
+•}•#•end•of•for:•access•each•word•class¶
```

```
>•sum.closed;•sum.open•#•look•at•the•output¶
[1]•913
[1]•2169
```

The following is already a much better way. It requires only two lines of code and avoids a loop to access all elements of Class:

```
>•sum(TokenFrequency[which(Class=="closed")])¶
[1]•913
>•sum(TokenFrequency[which(Class=="open")])¶
[1]•2169
```

But the best way involves a function called tapply. This function can take up to four arguments, but we will only look at the first three. The first is a vector to which you would like to apply a particular function. The second is a factor or a list of factors with as many elements as the vector to which you would like to apply the function. The third argument is the function you wish to apply, which can be one of many, many functions available in R (min, max, mean, sum, sd, var, . . .) as well as ones you define yourself:

```
>•tapply(TokenFrequency,•Class,•sum)¶
closed•••open
•••913•••2169
```

This is the way to tell R "for every level of Class, determine the sum of the values in TokenFrequency". Note again that the second argument can be a list of vectors/factors to look at more than one classifying factor at the same time.

You should now do "Exercise box 3.5: tapply" . . .

Another area where the apply family of functions becomes particularly useful is the handling of lists. We have seen above that lists are the most versatile data structure. We have also seen that subsetting by means of double square brackets allows us to access various parts of lists. However, there are other ways of getting information from lists that we have not dealt with so far. Let us first generate a list as an example:

```
>•another.list<-list(c(1),•c(2,•3),•c(4,•5,•6),•c(7,•8,•9,•10))¶
```

One useful thing to be able to do is to determine how long all the parts of a list are. With the knowledge we have now, we would probably proceed as follows:

```
>•lengths<-vector()¶
>•for•(i•in•1:length(another.list))•{¶
+••••lengths[i]<-length(another.list•[[i]])¶
+•}¶
>•lengths¶
[1]•1•2•3•4
```

However, with `sapply` this is much simpler. This function takes several arguments, some of which we will discuss here. The first argument is a list or a vector to whose elements you want to apply a function. The second argument is the function you wish to apply. The "s" in `sapply` stands for "simplify", which means that, if possible, R coerces the output into a vector format (or matrix) format. From this you may already guess what you can write:

```
>•sapply(another.list,•length)¶
[1]•1•2•3•4
```

This means "to every element of `another.list`, apply `length`". Another use of `sapply` allows you to access several parts of lists. For example, you may want to retrieve the first element of each of the four vectors of the list `another.list`. This is how you would do it with loops:

```
>•first.elements<-vector()¶
>•for•(i•in•1:length(another.list))•{¶
••••first.elements[i]<-another.list[[i]][1]¶
+•}¶
>•first.elements¶
[1]•1•2•4•7
```

The more elegant way uses subsetting as a function. That is to say, in the previous example, the function we applied to `another.list` was `length`. Now, we want to access a subset of elements, and we have seen above that subsetting is done with single square brackets. We therefore only need to use a single opening square bracket in place of `length` and then provide as a third argument the positional index, 1 in this case because we want the first element:

```
>•sapply(another.list,•"[",•1)¶
[1]•1•2•4•7
```

Note that this works even when the elements of the list are not all of the same kind:

```
>•sapply(a.list,•"[",•1)•#•use•the•list•we•generated•in•Section•3.5¶
[[1]]
[1]•1
```

```
$PartOfSpeech
[1] ADJ  ADV  N    CONJ PREP
Levels: ADJ ADV CONJ N PREP
[[3]]
[1] "This"
```

This last example already hints at something interesting. The function `sapply` and its sister function `lapply` apply to the data structure given in the first argument the function called in the second argument. Interestingly, the third argument provided is one that `sapply` and `lapply` just pass on to the function mentioned as the second argument. Let me give a simple example:

```
> a<-c(1, 5, 3); b<-c(2, 6, 4); (ab<-list(a, b))
[[1]]
[1] 1 5 3
[[2]]
[1] 2 6 4

> lapply(ab, sort, decreasing=F) # decreasing=F is passed on ; to sort
[[1]]
[1] 1 3 5
[[2]]
[1] 2 4 6
```

In a way, `lapply` takes the argument `decreasing=F` and passes it on to `sort`, which then uses this specification to sort the elements of the list `ab` in the specified way. These functions provide an elegant and powerful way of processing lists and other data structures. We will see applications of these uses of `tapply`, `sapply`, and `lapply` below.

The second useful programming guideline is to split up what you want to do into many smaller parts and make sure that you regularly save/output interim results—either into new data structures or even into files on your hard drive. While this may occasionally counteract your desire to let programs operate as quickly and memory-economically as possible—generating new data structures costs memory, saving interim results into files costs time—it is often worth the additional effort: first, if your program crashes, having split up your program into many smaller parts will make it easier for you to determine where and why the error occurred. Second, if your program crashes, having saved interim results into other data structures or files regularly will also decrease the amount of work you have to do again.

In this context, if your program involves one or more loops, it is often useful to build in a line that outputs a "progress report" to the screen so that you can always see whether your program is still working and, if it crashes, where the crash occurred. I will exemplify this below.

Finally, you will soon find that I sometimes use extremely long names for data structures and also make seemingly excessive use of comments. I realize these variable names are not particularly pleasant to type and may even increase the risk of typos, but this is still a useful practice. If you write code consisting of more than just a handful of lines and do not look at a particular script for a few days, you will notice how easily one forgets what exactly, say, a particular vector contains if you did not use comments and revealing names for data structures. Names such as `sentences.without`

`.tags` or `cleaned.corpus.file` are infinitely more revealing than the kind of names (e.g. x or aa) we used so far.

> For further study/exploration:
>
> - on how to apply a function to user-specified margins of data structures: ?apply¶
> - on how to apply a function elementwise to a data structure: ?lapply¶
> - on how to perform operations similar to tapply, but with a different output format and/or more flexibly: ?aggregate¶ and ?by¶

3.7 Character/String Processing

So far, we have only been concerned with generating and accessing (parts of) different data structures in R. However, for corpus linguists it is more important to know that R also offers a variety of sophisticated pattern matching tools for the processing of character strings. In this section, we will introduce and explore some of these tools before we apply them to corpus data.

There is one extremely important point to be made right here at the beginning: *know your corpora!* As you hopefully recall from the first exercise box in Section 2.1, a computer's understanding of what a word is may not coincide with yours, and that of any other linguist may also be very different from yours. It is absolutely imperative that you know exactly

- how you want to define what a word is (you may want to return to Exercise Box 2.1 for a moment to freshen up your memory), which may require quite a bit of thinking on your part before you begin writing your functions;
- what the corpus file looks like, which may require going over corpus documentation or, if you use unannotated texts, over parts of the corpus files themselves, in order to get to know spelling and annotation conventions.

A little example of something I came across in an undergrad corpus linguistics course was that we generated a concordance of *in* and found that in one of the files of the British National Corpus that we looked at, the word *in* was tagged as VBZ, which means "third person singular form of to be". All of us were stunned for a second, but then we noticed what was going on—can you guess?

THINK

BREAK

The answer is that there were instances of *innit* as a reduction of *isn't it* in the files, and *innit* is tagged in the British National Corpus as "<w·VBZ>in<w·XX0>n<w·PNP>it" . . . Thus, you really must know what's in your data.

You may *now* think, "But can't this be done more easily? Surely, ready-made programs wouldn't require me to do that?" Well, the answer is, "No! And wrong!" There is no alternative to knowing your corpora, this cannot be done more easily, and those concordance programs that come with more refined search options also require you to consider the above two questions thoroughly. Take, for example, a commercial program such as MonoConc Pro 2.2. True, you can generate the simplest concordances very easily, but don't get betrayed by the apparent simplicity. Once you become interested in more complex searches with tags, wildcards, etc., MonoConc Pro 2.2 offers a wide variety of settings—many more than some other concordance programs—and you will have to explore and set them on the basis of your needs and knowledge of the corpus files to get the most out of the program just like you need to set them in R. MonoConc Pro would also output *in* as a form of *to be* if that's what's in the corpus. Now, since, as a programming language, R is much more powerful than any ready-made concordancer, you must pinpoint what you want and you must familiarize yourself with the corpus carefully before you begin. This issue will surface repeatedly below in this section and throughout all of Chapter 4.

3.7.1 Getting Information from and Accessing (Vectors of) Character Strings

The most basic character operation in R gives the length of all character strings within a vector. The function nchar takes as its most important argument a character string (or the name of a vector containing character strings) and outputs the number of characters of the character string (or of all character strings in the vector):

```
>•example<-c("I",•"do",•"not",•"know")¶
>•nchar(example)¶
[1]•1•2•3•4
```

If you want to access a part of a character string, you can use substr. Most of the time, we will use substr with three arguments. The first is the character string (or the vector of character strings) that you want to access; the second one is the position of the first character you want to access; the third one is the position of the last character you want to access:

```
>•substr("internationalization",•6,•13)¶
[1]•"national"
```

or:

```
>•substr(example,•2,•3)¶
[1]•""•••"o"••"ot"•"no"
```

Note that substr handles vectors very efficiently:

```
>•some.first.vector<-c("abcd",•"efgh")¶
>•some.other.vector<-c("ijkl",•"mnop")¶
>•substr(c(some.first.vector,•some.other.vector),•c(1,•2,•3,•4),•
    c(2,•3,•4,•4))¶
[1]•"ab"•"fg"•"kl"•"p"
```

As you can see, first some.first.vector and some.other.vector are combined into one vector with four elements, and then the four different start and end positions of characters are applied to this character vector. This is going to be extremely useful below.

3.7.2 Elementary Ways to Change (Vectors of) Character Strings

The most basic ways in which (vectors of) character strings can be changed involves changing all characters to lower or upper case. The names of the relevant functions are `tolower` and `toupper` respectively, which both just take the (vectors of) character strings as their arguments.

```
>•tolower(example)¶
[1]•"i"••••"do"•••"not"••"know"
>•toupper(example)¶
[1]•"I"••••"DO"•••"NOT"••"KNOW"
```

Another elementary way to change (vectors of) character strings involves replacing x characters in (vectors of) character strings by x other characters. The name of the function is `chartr`, and it takes three arguments: first, the character(s) to be replaced; second, the character(s) that are substituted; third, the vector to which the operation is applied:

```
>•chartr("o",•"x",•example)¶
[1]•"I"••••"dx"•••"nxt"••"knxw"
```

This can be very useful, for example, to perform comprehensive transliteration operations in a single line of code.

3.7.3 Merging and Splitting (Vectors of) Character Strings without Regular Expressions

If you want to merge several character strings into one, you can use `paste`. These are the default arguments this function takes and their settings (where applicable):

- several different character strings or a vector of character strings;
- `sep="•"`: the character string(s) used to separate the different character strings when they are merged; the space is the default;
- `collapse=NULL`: the character string used to separate the different character strings when they are merged into just a single character string.

The following examples illustrate the use of this function and its arguments. If you want to paste together several vectors of length 1, `sep` and `collapse` don't make a difference:

```
>•paste("I",•"do",•"not",•"know",•sep="•")¶
[1]•"I•do•not•know"
>•paste("I",•"do",•"not",•"know",•collapse="•")•#•same•result¶
[1]•"I•do•not•know"
```

```
>•paste("I",•"do",•"not",•"know",•sep="•",•collapse="•")•#•same•result¶
[1]•"I•do•not•know"
```

On the other hand, if what is merged are vectors with more than one element, then `collapse` makes sure that all the vectors' elements get merged into a single character string:

```
>•paste(example,•sep="•")¶
[1]•"I"••••"do"•••"not"••"know"


#•with•a•longer•vector,•there•is•a•difference¶
>•paste(example,•sep="•",•collapse="•")•#•but•sep="•"•is•the•
   default•and•can•be•omitted¶
[1]•"I•do•not•know"


>•paste(example,•collapse="•")¶
[1]•"I•do•not•know"
```

By contrast, if you wish to split up a character string into several character strings, you can use `strsplit`. Its first argument is a character string to be split up; the second argument is the character string that `strsplit` will delete to (by means of which the first character string will be split up):

```
>•example.2<-"I•do•not•know"¶
>•strsplit(example.2,•"•")¶
[[1]]
[1]•"I"••••"do"•••"not"••"know"
```

Note that if you give an empty character string as the second argument, R will split up the first argument character-wise.

```
>•strsplit(example.2,•"")¶
[[1]]
•[1]•"I"•"•"•"d"•"o"•"•"•"n"•"o"•"t"•"•"•"k"•"n"•"o"•"w"
```

It is important to note that the output of `strsplit` is *not* a vector—it is a list of vectors (as you can infer from the double square brackets). The same works for larger vectors of character strings: the list has as many parts as `strsplit`'s first argument has parts:

```
>•example.3<-c("This•is•the•first•character•string",•"This•is•the•
   second•character•string")¶
>•strsplit(example.3,•"•")¶
[[1]]
```

```
[1]•"This"•••••••"is"••••••••"the"•••••••"first"••••••"character"•"string"
[[2]]
[1]•"This"•••••••"is"••••••••"the"•••••••"second"••••"character"•"string"
```

If you want to store all the separate character strings in one vector, just add `unlist`:

```
>•unlist(strsplit(example.3,•"•"))¶
[1]•"This"••••••"is"••••••••"the"•••••••"first"••••••"character"•"string"
[7]•"This"••••••"is"••••••••"the"•••••••"second"••••"character"•"string"
```

While these examples, in which character strings are split up on the basis of, say, just one space, are straightforward, `strsplit` is much more powerful because it can use regular expressions. However, before we introduce regular expressions, we will first treat some other functions used for searching and replacing character strings.

> For further study/exploration:
>
> - on how to extract/replace a part of a string: ?substr¶ (but cf. also below)
> - on how to trim character strings to user-specified widths: ?strtrim¶
> - on how to shorten strings to at least a user-defined length such that they remain unique: ?abbreviate¶

3.7.4 Searching and Replacing without Regular Expressions

In this section, we will introduce several pattern matching functions in a very basic manner, i.e. without using the powerful regular expression techniques involving character classes and place-holders/wildcards. To that end, let us first generate a vector of character strings:

```
>•text<-c("This•is•a•first•example•sentence.",•
  "And•this•is•a•second•example•sentence.")¶
```

The first function is called `grep`. If it is executed without invoking its regular expression capabilities, its most basic form requires only two arguments: a character string to search for and a vector of character strings (or something that can be coerced into a vector of character strings, such as a factor) in which the first character string is to be found. The function returns the positions of matches in the vector that was searched.

```
>•grep("second",•text)¶
[1]•2
```

This output illustrates two properties of `grep` you may not have expected: first, in this most basic form, `grep` does not return the match, but the position of the match. However, if you still remember what we said about subsetting above in Section 3.2.3, you should be able to figure out a way of how to arrive at the matching elements.

THINK

BREAK

```
>•text[grep("second",•text)]¶
[1]•"And•this•is•a•second•example•sentence."
```

However, this result will usually be produced slightly differently. Unless specified otherwise, `grep` assumes a default setting for a third argument, `value=F`, which means that only the positions are returned. But if you set `value=T`, you will get the matches instead of their positions.

```
>•grep("second",•text,•value=T)¶
[1]•"And•this•is•a•second•example•sentence."
```

Second, R does of course not know what a word is: it searches for character strings and returns the positions of the matches or the matches themselves irrespective of what is searched for corresponds to what a user considers a word or not.

```
>•grep("eco",•text,•value=T)¶
[1]•"And•this•is•a•second•example•sentence."
```

Third, `grep` returns the position of matches in the vector that was searched maximally once. The following line returns "1" and "2" only once each, although "is" occurs twice in both first lines, once in "This" or "this" and once in "is". Thus, `grep` just answers the question "Does something occur somewhere at all?", but not "does something occur somewhere at all and, if so, how often?"

```
>•grep("is",•text)¶
[1]•1•2
```

Finally, `grep` returns the positions of matches in the vector that was searched—e.g. the character string "second" occurs in the second character string of the vector `text`—it does *not* return the positions of the matches in the elements of the vector, i.e. the fact that the character string "second" begins at character position 15 in the second element of the vector `text`. However, this information can be retrieved with a different function: `regexpr`. The function `regexpr` takes the same two arguments as the most basic form of `grep`, but its output is different:

```
>•regexpr("second",•text)¶
[1]•-1•15
```

```
attr(,"match.length")
[1]•-1••6
```

The function returns two things. First, it returns a vector with the positions of the matches in the elements of the character vector (and instances of −1 for when there is no match). Thus, you can see that the character string "second" does not occur in the first character string of the vector `text`, but does occur in the second character string of the vector `text`, starting at position 15. Second, it returns a second vector that is an attribute of the first vector, which is called `match.length` and provides the lengths of the matches (and again −1 when there is no match). Obviously, there is no match for the character string `second` in the first character string of the vector `text`, but there is a match in the second character string of the vector `text` and it is six characters long. The lengths of the matches can be recovered for further processing in two ways. Either you call `attributes` with one argument—the element whose attributes you want as an argument—and get the lengths as a list . . .

```
>•attributes(regexpr("second",•text))•#•returns•a•list!¶
$match.length
[1]•-1••6
```

. . . or you call `attr` with two arguments—the element whose attributes you want as the first argument and the names of the attribute as the second argument—and get, more conveniently, a vector:

```
>•attr(regexpr("second",•text),•"match.length")•#•returns•a•vector!¶
[1]•-1••6
```

However, `regexpr` only outputs the starting position and length of the *first* possible match per element of the character vector:

```
>•regexpr("e",•text)¶
[1]•17•16
attr(,"match.length")
[1]•1•1
```

As of R 2.2, the additional function `gregexpr` (for "global `regexpr`") returns a list with the starting positions and lengths of *all* matches at the same time. This list has as many vectors as the data structure that was searched has vectors, and each of these vectors contains positive integers that correspond to the starting positions of all matches (and values of −1, if there is no match). Each of these vectors comes with a `match.length` attribute that provides the lengths of the matches. It is important that you understand this hierarchical arrangement: again, the list returned by `gregexpr` has vectors (one for each element of the vector that was searched), and each of the vectors in the list has an attribute called `match.length`:

```
>·gregexpr("e",·text)¶
[[1]]
[1]·17·23·26·29·32
```
← vector with starting positions of matches for the first element of text

```
attr(,"match.length")
[1]·1·1·1·1·1
[[2]]
[1]·16·22·28·31·34·37
```
← attributes of that vector with lengths of matches

← vector with starting positions of matches for the second element of text

```
attr(,"match.length")
[1]·1·1·1·1·1·1
```
← attributes of that vector with lengths of matches

It is possible to access both the starting points and the lengths of the matches in two different kinds of ways. The first of these preserves the information in which element of the searched vector the match was found, i.e. it gives you starting positions and lengths for each element of the searched vector; the second does not preserve this information and gives you all starting positions irrespective of where they were found.

Let us first look at how to retrieve the starting positions. We begin by retrieving the starting positions for each element of the searched vector. To that end, you can either use double square brackets (the attributes are listed but that doesn't matter, you can still use this like any other vector):

```
>•gregexpr("e",•text)[[1]]•#•and•of•course•the•same•with•[[2]]¶
[1]•17•23•26•29•32
attr(,"match.length")
[1]•1•1•1•1•1
```

... or you can use unlist to change the relevant part of the output of gregexpr into a vector. Why is that? This is because gregexpr as such returns a list, and we have seen above in Section 3.5 that if you access lists with single square brackets, you get a list; thus: unlist:

```
>•unlist(gregexpr("e",•text)[1])•#•and•of•course•the•same•with•[2]¶
[1]•17•23•26•29•32
```

Another way, which returns a list but gets rid of the attributes, involves sapply:

```
>•sapply(gregexpr("e",•text),•c)¶
[[1]]
[1]•17•23•26•29•32
[[2]]
[1]•16•22•28•31•34•37
```

The second way to access starting positions allows you to do so regardless of which element of the searched vector they occur in: you just do not subset:

```
>•unlist(gregexpr("e",•text))¶
[1]•17•23•26•29•32•16•22•28•31•34•37
```

Let us now turn to the lengths of the matches, which you can access in the same two ways as you accessed the starting points, preserving the information to which element the lengths belong and without preserving it. First we look at how to access the match.length attributes. While you will want to use attributes, note that this does not work:

```
>•attributes(gregexpr("e",•text))¶
NULL
>•attributes(gregexpr("e",•text)[1])¶
NULL
```

Why not?

THINK

BREAK

Because this line tries to retrieve attributes from the *list* returned by gregexpr, but I have told you that the attributes containing the lengths are not attributes of the list, but of the *vectors of the list.* Since you don't have vectors but a list, you must use the subsetting approach for lists that does not return lists but the data structures within the lists, i.e. double square brackets. For example, this is the first vector whose attributes you want:

```
>•gregexpr("e",•text)[[1]]¶
[1]•17•23•26•29•32
attr(,"match.length")
[1]•1•1•1•1•1
```

Thus, if you use double square brackets to get at the vector, then you can access its attributes with attributes to again get a list, which you could unlist...

```
>•attributes(gregexpr("e",•text)[[1]])•#•returns•a•list,•same•with•[[2]]¶
$match.length
[1]•1•1•1•1•1
>•unlist(attributes(gregexpr("e",•text)[[1]]))¶
match.length1•match.length2•match.length3
•••••••••••••1•••••••••••••1•••••••••••1
match.length4•match.length5
•••••••••••1•••••••••••1
```

... or again more conveniently with `attr` to immediately get a vector:

```
>•attr(gregexpr("e",•text)[[1]],•"match.length")•#•returns•a•vector,•
  same•with•[[2]]¶
[1]•1•1•1•1•1
```

Another possibility of retrieving the length for each element of the searched vector would be to retrieve sublists from the list returned by `gregexpr` (by writing `gregexpr("e",•text)[1]` and `gregexpr("e",•text)[2]`) and then to use `sapply` to apply either `attributes` or `attr` to each of the elements of the sublist, which you know to be the vector of starting positions of which you want the attributes. Sounds complex, but will in fact be obvious:

```
>•sapply(gregexpr("e",•text),•attributes)¶
$match.length
[1]•1•1•1•1•1
$match.length
[1]•1•1•1•1•1•1
```

... or ...:

```
>•sapply(gregexpr("e",•text),•attr,•"match.length")¶
[[1]]
[1]•1•1•1•1•1
[[2]]
[1]•1•1•1•1•1•1
```

Now, from this you can already infer that, if you want to access all the lengths of the matches at the same time regardless of which element of the searched vector they occur in, you just write this:

```
>•unlist(sapply(gregexpr("e",•text),•attr,•"match.length"))¶
[1]•1•1•1•1•1•1•1•1•1•1•1
```

Examples and practice will make you recognize how useful this all actually is. We will use `gregexpr` in many applications below.

The next function to be mentioned here is special. All functions dealt with thus far were part of the basic installation of R. The next function is called `strapply` and is part of Gabor Grothendieck's package `gsubfn` and offers functionality that is so useful for our corpus-linguistic purposes that it *must* be discussed here. It can take a variety of arguments, and we will discuss two of these here (and three more in the following section):

- a list or a vector to be searched for a particular pattern;
- a regular expression pattern to be retrieved in the list/vector.

If we use strapply in this minimal variant, it allows us to do searches in the way we know from text processing software, etc. in the sense that when we search for a character string, we do not get the vector elements with a match (as with grep), we do not get the vector elements together with the starting position(s) and lengths of the match(es) (as with gregexpr)—we get the match, and we get it in the form we know from gregexpr, namely as a list, each element of which contains the matches for the corresponding element of the character vector:

```
>•library(gsubfn)¶
>•text<-c("This•is•a•first•example•sentence.",•
  "And•this•is•a•second•example•sentence.")¶
>•strapply(text,•"first")¶
[[1]]
[1]•"first"
[[2]]
NULL

>•strapply(text,•"is")¶
[[1]]
[1]•"is"•"is"
[[2]]
[1]•"is"•"is"
```

While this is already quite convenient, the real power of this function will only become apparent in the following section.

All of these functions so far have only accessed (parts of) character vectors, but we have not changed the character vectors themselves yet. However, for many cases, this is of course exactly what we want to do. For example, we might want to delete all numbers and/or tags from a corpus file, we might want to insert tab stops before and after every match to render our results more easily handable in a spreadsheet program . . . In R, the function for substitutions is called sub. However, just like regexpr, sub only performs substitutions on the first match it finds, which is why we will immediately restrict our attention to the global substitution function gsub. The following are the main arguments of gsub and their default settings (again first without regular expressions, which will be introduced below):

```
gsub(pattern,•replacement,•x,•ignore.case=F)¶
```

The first argument is the expression that you want to replace, the second is the one you want to substitute instead. The third argument is the character vector you want to search and modify. The fourth argument specifies whether the search to be conducted will be case-sensitive (ignore .case=F) or not (ignore.case=T). Let us apply this function to our small character vector text. For example, we might want to change the indefinite determiner "a" to the definite determiner "the". By now, you should already know that the following will not really work well and why:

```
>•gsub("a",•"the",•text)¶
```

This will not work because with this function call we would assume that R 'knows' that we are only interested in changing "a" to "the" when "a" is a word, the indefinite determiner. But of course R does not know that:

```
>•gsub("a",•"the",•text)¶
[1]•"This•is•the•first•exthemple•sentence."
[2]•"And•this•is•the•second•exthemple•sentence."
```

Thus, we need to be more explicit. What is the most primitive way to do this?

```
>•gsub("•a•",•"•the•",•text)¶
[1]•"This•is•the•first•example•sentence."
[2]•"And•this•is•the•second•example•sentence."
```

Note, however, that R just outputs the result of gsub—R has not changed text:

```
>•text¶
[1]•"This•is•a•first•example•sentence."
[2]•"And•this•is•a•second•example•sentence."
```

Only if we assign the result of gsub to text again or some other vector will we be able to access the vector with its substitutions:

```
>•(text.2<-gsub("•a•",•"•the•",•text))¶
[1]•"This•is•the•first•example•sentence."
[2]•"And•this•is•the•second•example•sentence."
```

While all these examples are probably straightforward to the point of being self-evident, the real potential of the functions discussed above only emerges when they are coupled with regular expressions—i.e. placeholders/wildcards—to which we will turn now.

3.7.5 Searching and Replacing with Regular Expressions

In this section, we introduce the most important aspects of regular expressions, which will be the most central tool in most of the applications to be discussed below.[13] While R offers three types of regular expressions—basic, extended, and Perl-compatible—we will only be concerned with the Perl-compatible regular expressions because Perl is already widely used so chances are that you can apply what you learn here in different contexts. Thus, in order to maximally homogenize the explanations below, all regular expressions from now on will be written with the argument perl=T even if that may not be necessary in all cases.

Let us begin with expressions that do not consume characters but that specify (or "anchor") the position of character strings within character strings. The caret "^" and the "$" specify the beginning and the end of a character string. Thus, the following call only matches the first character string in the vector text since only this one begins with a "t" or, ignore.case=T, a "T":

```
>•grep("^t",•text,•ignore.case=T,•perl=T,•value=T)¶
[1]•"This•is•a•first•example•sentence."
```

Note that these expressions do not match any particular character in the string that is being searched: "^" does not match the first character of the line—it matches the beginning of the line *before* the first character: you can see that we still provided the "t" in the search string, which *is* the first character. Thus, these expressions are also referred to as zero-width tests.

The next expressions we deal with are not zero-width tests—they do match and consume characters. The period ".", for example, means "any single character except for the newline".[14] Thus, for example, if you want to find all occurrences of "s" that are followed by one other character and then a "c", you could enter the following into R:

```
>•grep("s.c",•text,•ignore.case=T,•perl=T,•value=T)¶
[1]•"And•this•is•a•second•example•sentence."
```

And of course, if you want a larger, fixed number of intervening characters, you just use as many periods as are necessary:

```
>•grep("f...t",•text,•ignore.case=T,•perl=T,•value=T)¶
[1]•"This•is•a•first•example•sentence."
```

But now this raises the question what to do when we want to look for a period, i.e. when we do not want to use the period as a metacharacter but literally? Obviously, we cannot take the period as such because the period stands for any one character but the newline. Thus, we need to know two things. First, a character that tells R "do not use a particular character's wildcard behavior but rather the literal character as such". These are called escape characters and in R this escape character is the backslash "\". Thus, you write:

```
>•gsub("\\.",•"!",•text,•perl=T)¶
[1]•"This•is•a•first•example•sentence!"
[2]•"And•this•is•a•second•example•sentence!"
```

Now, why do we need two backslashes? The answer is that the two backslashes are actually just one character, namely the escape character followed by an actual backslash (as you can easily verify by entering nchar("\\");•cat("\\")¶ into R). Thus, the two backslashes are in fact just one—the first one instructs R to treat the second as a literal character and not as an escape character for the period.

Second, we need to know which characters need to be escaped. The following is a list of such characters and their non-literal meanings, all of which will be used further below:

Line anchors:

^   'the beginning of a character string' (but cf. below)

$ 'marks the end of a character string'

Quantifying expressions:

*   'zero or more occurrences of the preceding regular expression'

+   'one or more occurrences of the preceding regular expression'

?   'zero or one occurrences of the preceding regular expression'

{ and }   ranges, mark the beginning and the end of a customizable quantifying expression

Other expressions:

·   'any one character but the newline'

\   the escape character

(and)   mark the beginning and the end of a (part of a) regular expression to be treated as a single element

| 'or'

[and]   mark the beginning and the end of a character class

"   a double quote needs to be escaped if you use double quotes to delimit your character vector that makes up the regular expression

'   a single quote needs to be escaped if you use single quotes to delimit your character vector that makes up the regular expression (which we don't do)

While the period stands for any one character apart from the newline, you may want to do more than retrieve exactly one kind of match. The above list already shows some of the quantifying expressions you can use. For example, if you have a vector in which both the British and the American spellings of *colour* and *color* occur and you would like to retrieve both spelling variants, you can make use of the fact that a question mark means 'zero or one occurrence of the immediately preceding expression'. Thus, you can proceed as follows:

```
>•colors<-c("color",•"colour")¶
>•grep("colou?r",•colors,•perl=T,•value=T)¶
[1]•"color"••"colour"
```

Another domain of application of such quantifying expressions would be to clean up text files by, for example, changing all sequences of more than one space to just one space. How would you do this?

 **THINK**

BREAK

Well, you could write:

```
>•some.text.line<-"This••is••••just•one••••example."¶
>•gsub("•+",•"•",•some.text.line,•perl=T)¶
[1]•"This•is•just•one•example."
```

As a matter of fact, using the question mark and the plus sign are just two special shortcuts for a more general notation that allows you to specify, for example,

- the exact number of matches *m* of an expression you would like to match: just add {m} to an expression;
- the minimum number of matches *m* of an expression you would like to match (by leaving open the maximum): just add {m,} to an expression;
- the minimum number of matches *m* and the maximum number of matches *n* of an expression you would like to match: just add {m,n} to an expression.

Thus, the application of the question mark above is actually just a shorthand for an expression involving ranges. So, how could you rewrite grep("colou?r",•colors,•perl=T,•value=T)¶ using a (in this case clumsy) range expression?

THINK

BREAK

This would be it:

```
>•grep("colou{0,1}r",•colors,•perl=T,•value=T)¶
[1]•"color"••"colour"
```

It is important here to clarify the range to which these quantifying expressions apply. From the previous examples, it emerges that the quantifiers always apply only to the immediately preceding element. However, sometimes you may want to quantify something that consists of more than one element. The way to do this in R has already been hinted at above: you use opening and closing parentheses to group elements (characters and/or metacharacters). Thus, for example, if you want to match two or three occurrences of "st", you could do this:

```
>•some.vector<-c("a•st•b",•"a•stst•b",•"a•st•st•b")¶
>•grep("(st){2,3}",•some.vector,•perl=T,•value=T)¶
[1]•"a•stst•b"
```

As you can see, the expression in parentheses is treated as single element, which is then quantified as specified. Of course, such an expression within parentheses can itself contain metacharacters, further parentheses, etc.

```
>•grep("(.t){2,3}",•some.vector,•perl=T,•value=T)¶
[1]•"a•stst•b"
```

In fact, one of the probably more frequent domains in which the notation with parentheses is applied involves providing R with several alternative expressions to be matched. To do that, you simply put all alternatives in parentheses and separate them with a vertical bar (or pipe) |.

```
>•gsub("a•(first|second)",•"another",•text,•ignore.case=T,•perl=T)¶
[1]•"This•is•another•example•sentence."
[2]•"And•this•is•another•example•sentence."
```

As another example, imagine that you want to replace all occurrences of "a", "e", "i", "o", and "u" by a "V" (for vowel). You could do this as follows:

```
>•gsub("(a|e|i|o|u)",•"V",•text,•ignore.case=T,•perl=T)¶
[1]•"ThVs•Vs•V•fVrst•VxVmplV•sVntVncV."
[2]•"Vnd•thVs•Vs•V•sVcVnd•VxVmplV•sVntVncV."
```

But since this kind of application—the situation where one wishes to replace one out of several individual characters with something else—is a particularly frequent one, there is actually a shorter alternative notation which uses square brackets without "|" to define a class of individual characters and replace them all in one go:

```
>•gsub("[aeiou]",•"V",•text,•ignore.case=T,•perl=T)¶
[1]•"ThVs•Vs•V•fVrst•VxVmplV•sVntVncV."
[2]•"Vnd•thVs•Vs•V•sVcVnd•VxVmplV•sVntVncV."
```

It is also possible to define such character classes in square brackets using the minus sign as range operator. For example, if you want to replace any occurrence of the letters "a" to "h" and "t" to "z" by an "X", you can do this as follows:

```
>•gsub("[a-ht-z]",•"X",•text,•ignore.case=T,•perl=T)¶
[1]•"XXis•is•X•XirsX•XXXmplX•sXnXXnXX."
[2]•"XnX•XXis•is•X•sXXonX•XXXmplX•sXnXXnXX."
```

Note that the minus sign is now not used as part of a character class: "[a-z]" does not mean 'match either "a" or "-" or "z"—it means 'match any of the letters from "a" to "z" '. If you want to include the minus sign in your character class, you can put it at the first position within the square brackets. Thus, to match either "a", "-", or "z", you write "[-az]" or "[-za]".

The same pattern works for numbers. For example, you could replace the numbers from 1 to 5 in the string "0123456789" with "3" as follows:

```
>•gsub("[1-5]",•"3",•"0123456789",•perl=T)¶
[1]•"0333336789"
```

You should now do "Exercise box 3.6: A few regular expressions" . . .

For cases in which you would like to match every character that does not belong to a character class, you can use the caret "^" as the first character within square brackets.

```
>•gsub("[^1-5]",•"3",•"0123456789",•perl=T)¶
[1]•"3123453333"
```

Note how the caret within the definition of a character class in square brackets means something else (namely, "not") than the caret outside of a character class (where it means "at the beginning of a string").

While it is of course possible to define all alphanumeric characters or all numbers as character classes in this way ("[a-zA-Z0-9]" and "[0-9]" respectively), R offers a few predefined character classes. One such character class, written as "\\d" in R (cf. above on double backslashes), is equivalent to "[0-9]", i.e. all digits:

```
>•gsub("\\d",•"3",•"a1b2c3d4e5",•perl=T)¶
[1]•"a3b3c3d3e3"
```

The opposite of this character class—i.e. what one might want to write as "[^\\d]"—can be expressed as "\\D":

```
>•gsub("\\D",•"3",•"a1b2c3d4e5",•perl=T)¶
[1]•"3132333435"
```

Another predefined character class that will turn out to be very useful is that of word characters, written as "\\w" in R. This character class is the short form of what you could define more lengthily as "[0-9A-Za-z_]", and again "\\W" is its opposite. From what you know by now, it should be obvious that you can combine this character class with a plus sign to look for what will, in the case of the Latin alphabet, correspond to words, at least words without apostrophes and hyphens in them:

```
>•gsub("\\w+",•"WRD",•text,•perl=T)¶
[1]•"WRD•WRD•WRD•WRD•WRD•WRD."•••••"WRD•WRD•WRD•WRD•WRD•WRD•WRD."
```

We will use this character class frequently in the next chapter. Then, there is the class written as "\\s", which stands for all whitespace characters: tab stops (otherwise written as "\t"), spaces (otherwise written as " "), newlines (otherwise written as "\n"), and carriage returns (otherwise written as "\r"). As before, changing "\\s" into "\\S" gives you the opposite character class.

The final metacharacter to be mentioned here is again an anchor, a zero-width test (just like the line anchors "^" and "$" mentioned above). The expression "\\b" refers to a word boundary, i.e. the position between a word character and a non-word character (in either order); again "\\B" is the opposite. That is to say, "\\b" is the position between "\\w\\W" or "\\W\\w", but includes neither "\\w" nor "\\W"! Note the differences between the first two searches and the third one using the zero-width pattern:

```
> gsub("\\w\\w", "<WB>", text, perl=T) # consumes the word character¶
[1] "Thi<WB>i<WB><WB>firs<WB>exampl<WB>sentenc<WB>"
[2] "An<WB>thi<WB>i<WB><WB>secon<WB>exampl<WB>sentenc<WB>"

> gsub("\\W\\w", "<WB>", text, perl=T) # consumes the word character¶
[1] "This<WB>s<WB><WB>irst<WB>xample<WB>entence."
[2] "And<WB>his<WB>s<WB><WB>econd<WB>xample<WB>entence."

> gsub("\\b", "<WB>", text, perl=TRUE) # does not consume anything¶
[1] "<WB>This<WB> <WB>is<WB> <WB>a<WB> <WB>first<WB> <WB>example
   <WB> <WB>sentence<WB>."
[2] "<WB>And<WB> <WB>this<WB> <WB>is<WB> <WB>a<WB> <WB>second
   <WB> <WB>example<WB> <WB>sentence<WB>."
```

There are two important potential problems that we have so far simply glossed over. The first of these is concerned with what R does when there are several ways to match a pattern. Let's try it out:

```
> (the.match<-gregexpr("s.*s", text[1], perl=T))¶
[[1]]
[1] 4
attr(,"match.length")
[1] 22
```

The answer is clear. We ask R to match any "s" followed by zero or more occurrences of any character followed by another "s". There would be many ways to match this in our vector. In the following lines, the possible matches for the first string of the vector text are underlined, and the first two lines give the number of the position within the character string:

```
00000000011111111112222222222333
12345678901234567890123456789 0123
This·is·a·first·example·sentence.
This·is·a·first·example·sentence.
This·is·a·first·example·sentence.
This·is·a·first·example·sentence.
This·is·a·first·example·sentence.
This·is·a·first·example·sentence.
```

Obviously, R prefers the third solution: it begins to match at the first occurrence of "s"—at position 4—and then chooses the longest possible match (if you do not understand this line yet, that's okay, this is just a primer and we'll come back to this in detail in Section 4.1.5):

```
>•substr(text[1],•unlist(the.match),•unlist(the.match)+attr(the.match[[1]],•
  "match.length")-1)¶
[1]•"s•is•a•first•example•s"
```

In regular expressions lingo, this is called greedy matching. However, this may not always be what we want. If, for example, we have a corpus line that matches a particular search pattern more than once, then, as corpus linguists, we probably would not just want one long match (with unwanted material in the middle).

```
>•some.corpus.line<-"he•said,•you•are•lazy•and•you•are•stupid."¶
>•gregexpr("you.*are",•some.corpus.line,•perl=T)¶
[[1]]
[1]•10
attr(,"match.length")
[1]•24
```

Rather, we are more likely to be interested in getting to know that there are two matches and what these are. There are two ways of setting R's matching strategy to lazy, i.e. non-greedy. The first one applies to a complete search expression, setting all quantifiers in the regular expression to non-greedy matching. You just write "(?U)" as the first four characters of your regular expression:

```
>•gregexpr("(?U)you.*are",•some.corpus.line,•perl=T)¶
[[1]]
[1]•10•27
attr(,"match.length")
[1]•7•7
```

As you can see, now both consecutive matches are found. Note in passing that there are some other useful switches that are used the same way, but the only other one we introduce here is "(?i)", which instructs R to perform a case-insensitive match (just as ignore.case=T did above). The second way to set R's matching to non-greedy is more flexible: rather than using non-greedy matching for the complete regular expression, you can use greedy *and* non-greedy matching in one expression by adding a question mark to only those quantifiers which you would like to use non-greedily. As you can see, we now get the desired result—both matches and their correct lengths are matched:

```
>•gregexpr("you.*?are",•some.corpus.line,•perl=T)¶
[[1]]
[1]•10•27
attr(,"match.length")
[1]•7•7
```

Note, however, what happens if we apply this solution to the vector text:

```
>·gregexpr("s.*?s",·text[1],·perl=T)¶
[[1]]
[1]··4·14
attr(,"match.length")
[1]··4·12
```

We still don't get all three consecutive occurrences we would probably want, namely "s·is", "s·a·firs", and "st·example·s". Why's that?

THINK

BREAK

This is because when the regex engine has identified the first match—"s·is"—then the second "s" has already been consumed as being the second "s" in the match and is, thus, 'not available' anymore to be the first match the next time. We will therefore revisit this case below.

The second issue we need to address is that we have so far only been concerned with the most elementary replacement operations. That is to say, while you have used simple matches as well as more complex matches in terms of what you were searching for, your replacement has so far always been a specific element (a character or a character string). However, there are of course applications where, for example, you may want to add something to our match. Imagine, for example, you would like to tag the vector text such that every word is followed by "<w>" and every punctuation mark is followed by "<p>". Now, obviously you cannot simply replace the pattern match as such with the tag because then your words would get replaced:

```
>·(text.2<-gsub("\\w+",·"<w>",·text,·perl=T))¶
[1]·"<w>·<w>·<w>·<w>·<w>·<w>."
[2]·"<w>·<w>·<w>·<w>·<w>·<w>·<w>."
```

Secondly, you cannot simply replace word boundaries "\\b" with tags because then you get tags before *and* after the words:

```
>·(text.2<-gsub("\\b",·"<w>",·text,·perl=T))¶
[1]·"<w>This<w>·<w>is<w>·<w>a<w>·<w>first<w>·<w>example<w>·<w>sentence<w>."
[2]·"<w>And<w>·<w>this<w>·<w>is<w>·<w>a<w>·<w>second<w>·<w>example<w>·
  <w>sentence<w>."
```

Thus, what you need is a way to tell R to (i) match something, (ii) bear in mind what was matched, and (iii) replace the match by what was matched *together with* the additional tag you want to insert. Fortunately, regular expressions in R (Perl, Python, and . . .) offer the possibility of what is referred to as "back-referencing". Regular expressions or parts of regular expressions that are grouped by parentheses are internally numbered (in the order of opening parentheses) so that they are available for future matching and/or replacing in the same function call. Let us first look at the simple tagging example from above. In two steps, you can tag the corpus in the above way.

```
>•(text.2<-gsub("(\\w+)",•"\\1<w>",•text,•perl=T))¶
[1]•"This<w>•is<w>•a<w>•first<w>•example<w>•sentence<w>."
[2]•"And<w>•this<w>•is<w>•a<w>•second<w>•example<w>•sentence<w>."
>•(text.3<-gsub("([!:,\\.\\?])",•"\\1<p>",•text.2,•perl=T))¶
[1]•"This<w>•is<w>•a<w>•first<w>•example<w>•sentence<w>.<p>"
[2]•"And<w>•this<w>•is<w>•a<w>•second<w>•example<w>•sentence<w>.<p>"
```

That is to say, in the first step, the pattern you are searching for is defined as one or more consecutive occurrences of a word character, and the parentheses instruct R to treat the complete sequence as a single element. Since there is only one set of opening and closing parentheses, the number under which the match is available is of course 1. The replacement part, "\\1<w>", then instructs R to insert the remembered match of the first parenthesized expression in the search pattern ("\\1")—which in this case amounts to the whole search pattern—and insert "<w>" afterwards. The second step works in the same way except that, now, the parenthesized subexpression is a character class that is treated as a single item, to which a punctuation mark tag "<p>" is added.

Of course, this approach can be extended to handle more complex expressions, too. Imagine, for example, your data contain dates in the American English format with the month preceding the day (i.e. Christmas would be written as "12/25/2006") and you want to reverse the order of the month and the day so that your data could be merged with corpus files that already use this ordering. Imagine also that some of the American dates do not use slashes to separate the month, the date, and the year, but periods. And, you must take into consideration that some people write April 1st as "04/01" whereas others write "4/1". How could you change American dates into British dates using slashes as separators?

```
>•American.dates<-c("5/15/1976",•"2.15.1970",•"1.9.2006")¶
```

THINK

BREAK

```
>•(British.dates<-sub("(\\d{1,2})\\D(\\d{1,2})\\D",•"\\2/\\1/",•
   American.dates,•perl=T))¶
[1]•"15/5/1976"•"15/2/1970"•"9/1/2006"
```

This regular expression may look pretty daunting at first, but in fact it is a mere concatenation of many simple things we have already covered. The first parenthesized expression, "(\\d{1,2})", matches one or two digits and, if matched, stores them as the first matched regular expression, "\\1". "\\D" then matches any non-digit character and, thus, both the period and the slash. Of course, if you are sure that the only characters in the data used to separate the parts of the date are a period and slash, a character class such as "[\\./]" would have worked just as well as "\\D". If you are not, then "\\D" may be a better choice because it would also match if the parts of the date were separated by hyphens as in "5-15-1976". The second parenthesized expression, again "(\\d{1,2})", matches one or two digits but, if matched, stores the match as the second matched regular expression, "\\2". You might actually want to think about how you can also preserve the original separators instead of changing them into slashes . . .

THINK

BREAK

```
>•(British.dates<-sub("(\\d{1,2})(\\D)(\\d{1,2})(\\D)",•"\\3\\4\\1\\2",•
  American.dates,•perl=T))¶
[1]•"15/5/1976"•"15.2.1970"•"9.1.2006"
```

So far we have used back-referencing in such a way that the match of a search expression or parts of the match of a search expression were used again in the *replacement pattern*. Note, however, that the above characterization of back-referencing also includes the possibility of reusing a match in the *search pattern*. Let us look at a simple example first. Imagine you are interested in the rhyming of adjacent words and would, therefore, like to find in a character vector just like text cases where adjacent words end with the same character sequences, obviously because these are likely candidates for rhymes.[15] More specifically, imagine you would like to add "<r>" to every potential rhyme sequence such that "This is not my dog" gets changed into "This<r> is<r> not my dog", but that you only want cases of adjacent words so that the string "This not is my dog" would not match. In such a case, you would want to find a sequence of characters at the end of a word, which we can translate as "\\w+?\\W+". However, it is more complex than that. You must instruct R to bear the sequence of characters in mind, which is why you write "(\\w+?)". Note that we use the + after "\\w" because we want at least one match. Then, this character sequence in question must be followed by at least one non-word character (e.g. a space or a comma and a space) plus the word characters of the beginning of the next word: "\\W+\\w*?". Note, however, that we write "\\w*?" because the beginning of the word may already rhyme so the beginning is optional, hence the asterisk. Since you will have to put those back into the string, you will need to memorize and, thus, parenthesize them, too. Then, you only want matches if the same character sequence you found at the beginning is matched again: "\\1". However, you only want it to match if this is again at the end of a word (i.e. followed by "\\W") and, what is more, you will have to memorize that non-word character to put it back in: "\\1\\W". As you can see, the back-reference is back to a pattern within the search.

When this is all done, the replacement is easy: you want to replace all this by the potentially rhyming characters followed by "<r>", then the characters after the potentially rhyming characters, then the potentially rhyming characters, and the next non-word character. All in all, this is what it looks like:

```
>•gsub("(\\w+?)(\\W+\\w*?)\\1(\\w)",•"\\1<r>\\2\\1<r>\\3",•text,•perl=T)¶
[1]•"This<r>•is<r>•a•first•example<r>•sentence<r>."
[2]•"And•this<r>•is<r>•a•second•example<r>•sentence<r>."
```

You can also test that this regular expression does not match *This not is my dog* because the *This* and the *is* are not adjacent anymore:

```
>•gsub("(\\w+?)(\\W+\\w*?)\\1(\\w)",•"\\1<r>\\2\\1<r>\\3",•
  "This•not•is•my•dog",•perl=T)¶
[1]•"This•not•is•my•dog"
```

Hopefully, you look at this and immediately see a way to improve it. If not, think about (or try out) what happens if the elements of the vector `text` did not have final periods. How could you improve the script in this regard and simplify it at the same time?

THINK

BREAK

The solution and simpler strategy would be this:

```
>•gsub("(\\w+?)(\\W+\\w*?)\\1\\b",•"\\1<r>\\2\\1<r>",•text,•perl=T)¶
```

As you can see, instead of the final non-word character, this uses a word boundary at the end. Why? This is because the expression "\\1\\b" only consumes the rhyming character(s), which has two consequences. First, this expression also matches the end of the string if there is no period because it does not require a separate non-word character to match after the rhyming character(s). Second, because the expression does not consume anything after the rhyming character(s), you need not capture anything after their second occurrence because you do not have to put anything back in. Thus, the second version is actually preferable.

This example and its discussion gloss over the additional issue of how the approach would have to be changed to find rhyming words that are not adjacent, and you will meet this issue in the next exercise box.

The penultimate kind of expression to be discussed here is referred to as lookaround. "Lookaround" is a cover term for four different constructs: positive and negative lookahead and positive and negative lookbehind. While lookaround is sometimes exceptionally useful, we will only discuss lookahead here, and only cursorily so, because while lookaround is one of the most powerful constructs, it can be very complex to use.

The key characteristics of these lookaround expressions are that:

- like most regular expressions we have seen before, they match characters (rather than, say, positions);
- unlike most regular expressions we have seen before, they give up the match—i.e. they do not consume characters and are therefore, just like line anchors, zero-width assertions—and only return whether they matched or did not match.

Let us look at a simple example for positive lookahead. You use positive lookahead to match something that is followed by something else (which, however, you do *not* wish to consume with that expression). The syntax for positive lookahead is to put the positive lookahead between regular parentheses and have a question mark and an equals sign after the opening bracket. For example, imagine you have the following vector `example`:

```
>•example<-c("abcd",•"abcde",•"abcdf")¶
```

If you now want to match "abc" but only if "abc" is followed by "de", you can write this:

```
>•grep("abc(?=de)",•example,•perl=T)¶
[1]•2
```

You may now ask, "What's the point, I could just write this?":

```
>•grep("abcde",•example,•perl=T)¶
[1]•2
```

True, but one area where the difference is particularly relevant is when you want to not just match, but also replace. If you want to replace "abc" by "xyz", but only if "abc" is followed by "de", you can't just write the following because it will also replace the "de":

```
>•gsub("abcde",•"xyz",•example,•perl=T)¶
[1]•"abcd"••"xyz"•••"abcdf"
```

With positive lookahead, no problem:

```
>•gsub("abc(?=de)",•"xyz",•example,•perl=T)¶
[1]•"abcd"••"xyzde"•"abcdf"
```

As you can see, the string "de" is not consumed, and thus not replaced. Of course, you can combine such lookaround expressions with all the regular expressions you already know. For example, you could replace every character in front of an "e" by two occurrences of that character:

```
>•gsub("(.)(?=e)",•"\\1\\1",•example,•perl=T)¶
[1]•"abcd"•••"abcdde"•"abcdf"
```

However, positive lookahead is not only useful for replacements. Let us now finally discuss how we deal with the problem of finding the three instances of one "s" followed by some characters up to the next that we faced with the vector text:

```
>•text¶
[1]•"This•is•a•first•example•sentence."
[2]•"And•this•is•a•second•example•sentence."
```

The last time we managed to get two out of the three desired matches (when we introduced lazy matching), but we still had the problem that a consumed second "s" was not available as a first "s" in the next match. As you may already guess, the answer is not to consume the second "s", and you now know how to do that—positive lookahead:

```
>•gregexpr("s.*?(?=s)",•text[1],•perl=T)¶
[[1]]
[1]••4••7•14
attr(,"match.length")
[1]••3••7•11
```

Finally . . . all three matches are identified.

Negative lookahead allows you to match something if it is not followed by something else. In this example, you capitalize only those "d" characters that are not followed by "e":

```
>•gsub("d(?!e)",•"D",•example,•perl=T)¶
[1]•"abcD"••"abcde"•"abcDf"
```

Note the difference from what you may think is the same, but which in fact is *not* the same:

```
>•gsub("d[^e]",•"D",•example,•perl=T)¶
[1]•"abcd"••"abcde"•"abcD"
```

Negative lookahead is especially useful when you want to replace unwanted strings, and here is an example we need to discuss in detail. Imagine you have a sentence with an SGML word-class annotation as in the British National Corpus (BNC). In this format, every word is preceded by a tag in angular brackets (cf. again the Appendix for a link to a website with all BNC tags, which you may want to download now because we will need it often). The first character after the opening angular bracket is the tag's name "w" (for "word"), which is mostly followed by a space and a three-character part-of-speech tag. However, the BNC also has so-called portmanteau tags, which indicate that the automatic tagger could not really decide which of two tags to apply. For example, a portmanteau tag "<w•AJ0-VVN>" would mean that the tagger was undecided between the base form of a regular adjective ("AJ0") or the past participle of a full lexical verb ("VVN"). Apart from POS tags, other information is also included. For example, tags named "s" contain a name-value pair providing the number of each sentence unit in a file; tags named "ptr" are used to indicate overlapping speech and contain a name-value pair identifying the location and the speaker of the overlapping speech. This is an example:

```
>•example1<-"<w•UNC>er<c•PUN>,•<w•AV0>anyway•<w•PNP>we<w•VBB>'re•
  <w•AJ0>alright•<w•AV0>now•<ptr•target=KB0LC003><w•AV0>so
  <c•PUN>,•<w•PNP>you•<w•VVB>know•<ptr•target=KB0LC004></u>"¶
```

Imagine now you want to delete all tags that are not word tags or punctuation mark tags. In other words, you want to keep all tags that look like this: "<[wc]·...(-....)?>". Unfortunately, you cannot do this with negated character classes: something involving "[^wc·]" does not work because R will understand the space as an alternative to "w" and "c", but not as meaning 'not "w·" and also not "c·"'. Now you might say, okay, but I can do this":

```
>•gsub("<[^wc][^•].*?>",•"",•example1,•perl=T)¶
[1]•"<w•UNC>er<c•PUN>,•<w•AV0>anyway•<w•PNP>we<w•VBB>'re•
  <w•AJ0>alright•<w•AV0>now•<w•AV0>so<c•PUN>,•<w•PNP>you•
  <w•VVB>know•"
```

And yes, this looks nice. But ... you are just lucky because example1 does not have a tag like "<wtr·target=KB0LC004>", which could exist and which you then would also want to delete. Look what would happen:

```
>•example2<-"<w•UNC>er<c•PUN>,•<w•AV0>anyway•<w•PNP>we<w•VBB>'re•
  <w•AJ0>alright•<w•AV0>now•<ptr•target=KB0LC003><w•AV0>so
  <c•PUN>,•<w•PNP>you•<w•VVB>know•<wtr•target=KB0LC004></u>"¶
>•gsub("<[^wc][^•].*?>",•"",•example2,•perl=T)¶
[1]•"<w•UNC>er<c•PUN>,•<w•AV0>anyway•<w•PNP>we<w•VBB>'re•
  <w•AJ0>alright•<w•AV0>now•<w•AV0>so<c•PUN>,•<w•PNP>you•
  <w•VVB>know•<wtr•target=KB0LC004>"
```

The last tag, which you wanted to delete, was not deleted. Why? Because it does not match: you were looking for something that is not "w" or "c" followed by something that is not a space, but this tag has a "w" followed by a non-space. The same problem arises when there is no "w" at the beginning of the tag in question, but a space in its second position:

```
>•example3<-"<w•UNC>er<c•PUN>,•<w•AV0>anyway•<w•PNP>we<w•VBB>'re•
  <w•AJ0>alright•<w•AV0>now•<ptr•target=KB0LC003><w•AV0>so
  <c•PUN>,•<w•PNP>you•<w•VVB>know•<p•tr•target=KB0LC004>
  </u>"¶
>•gsub("<[^wc][^•].*?>",•"",•example3,•perl=T)¶
[1]•"<w•UNC>er<c•PUN>,•<w•AV0>anyway•<w•PNP>we<w•VBB>'re•
  <w•AJ0>alright•<w•AV0>now•<w•AV0>so<c•PUN>,•<w•PNP>you•
  <w•VVB>know•<p•tr•target=KB0LC004>"
```

The answer to our problem is negative lookahead. Here's how you do it, and you can try it out with example2 and example3 as well to make sure it works:

```
>•gsub("<(?![wc]•...(-...)?>).*?>",•"",•example1,•perl=T)¶
```

It is important that you understand how this works so let me go over it in detail. First, the expression matches an opening angular bracket "<", no problem here. Then, negative lookahead says, match only if the parenthesized lookahead expression, i.e. the characters after the "?!" after the opening angular bracket, does not (*negative* lookahead) match "<[wc]· . . . (-. . .)?>". Now comes the tricky part. You must recall that lookahead does not consume: thus, after having seen that the characters after the angular bracket are not "[wc]· . . . (-. . .)?>", R's regular expression engine is still at the position after the first consumed character, the angular bracket; that's what zero-width assertion was all about. Therefore, after having said what you *don't* want to see after the "<"—which is "[wc]· . . . (-. . .)?>"—you must now say what you *do* want to see, which is anything, hence the ".*?" but only until the next angular bracket closing the tag. Another way to understand this is as follows: negative lookahead says "match the stuff after the negative lookahead (".*?"), but not if it looks like the stuff in the negative lookahead ("[wc]· . . . (-. . .)?>").

So how does it apply to example1? The first tag "<w·UNC>" is not matched because the expression cannot match the negative lookahead: "<w·UNC>" is exactly what "?!" says *not* to match. The same holds for the next few tags. Once the regex engine arrives at "<ptr·target=K B0LC003>", though, something different happens. The regex engine sees the "<" and notices that the characters that follow the "<" are not "[wc]· . . . (-. . .)?>". Thus, it matches, but does not consume; the engine is still after the "<". The string "ptr target=KB0LC003" is then matched by ".*?", and the closing ">" is matched by ">". Thus, everything matches, and R can replace this by " ", effectively deleting it.

This is a very useful approach to know and most of the scripts below that handle BNC files will make use of this expression to "clean up" the file before further searches take place. An extension of this does not only delete unwanted tags, but also the material that follows these unwanted tags up until the next tag:

```
> gsub("<(?![wc]· . . . (-. . .)?>).*?>[^<]*", " ", example1, perl=T)¶
```

For further study/exploration:

- if you use parentheses but don't want the characters in parentheses to be remembered, you can use (?: . . .)
- on how to match with positive and negative lookbehind: explore (?<=) and (?<!)

Finally, we return to Gabor Grothendieck's function strapply again. Above we only looked at two of its arguments, but now that we covered quite a few regular expressions, we can also get to know some more arguments. These are some of strapply's arguments; for the sake of convenience, I again include the two we already mentioned:

```
strapply(X, pattern, FUN, backref= . . . , perl=T, ignore.case=T)
```

- X is a list or a vector to be searched for a particular pattern;
- pattern: a regular expression pattern to be retrieved in the list/vector;
- FUN: an optional function which is applied to the matches and all back-referenced strings in the list/vector and whose result is outputted; if no function is provided as an argument,

strapply just returns the matches and all back-referenced strings as a list; this will be explained below;

- an optional backref=x, where x is an integer, an argument to which we will also turn below;
- perl=T or perl=F (we will only use the former) and also ignore.case=T or ignore.case=F.

This probably sounds a little cryptic so let us look at an example. We again use the vector text as an example:

```
>•text<-c("This•is•a•first•example•sentence.",•
  "And•this•is•a•second•example•sentence.")¶
```

We have seen what strapply can do for us—we get the exact match:

```
>•strapply(text[1],•"first")¶
[[1]]
[1]•"first"
```

However, the function has more to offer. Let's assume you want to know all the words that contain an "i" in the character vectors under consideration. You can then enter the following:

```
>•strapply(text,•"([a-z]*)i([a-z]*)",•ignore.case=T,•perl=T)¶
[[1]]
[1]•"This"••"is"••••"f"••••"rst"
[[2]]
[1]•"this"•"is"
```

Still no big deal, but now strapply's added bonus is that, as you may already have guessed from the "apply" part of the function's name, that you can add as an argument a function which is then applied to the matches. Thus, for example, retrieving the lengths of all words that contain an "i" in the vector text is really simple:

```
>•strapply(text,•"[a-z]*i[a-z]*",•nchar,•ignore.case=T,•perl=T)¶
[[1]]
[1]•4•2•5
[[2]]
[1]•4•2
```

Let us now look briefly at the argument backref. There are basically three possibilities: (i) if the search pattern includes backreferences but strapply has no backref argument, the function call returns *not* the complete match, but only the contents of the backreferences. (ii) If the search pattern includes backreferences and strapply has a negative number *n* as the backref argument, the function call returns *not* the complete match, but only the contents of the first *n* backreferences. (iii) If the search pattern includes backreferences and strapply has a positive number *n* as the backref argument, the function call returns the complete match and the contents of the first *n* backreferences:

```
>•strapply(text,•"([a-z]*)i([a-z]*)",•c,•ignore.case=T,•perl=T)¶
[[1]]
[1]•"Th"••"s"•••""••••"s"•••"f"•••"rst"
[[2]]
[1]•"th"•"s"••""•••"s"
```

```
>•strapply(text,•"([a-z]*)i([a-z]*)",•c,•ignore.case=T,•perl=T,•backref=-1)¶
[[1]]
[1]•"Th"•""•••"f"
[[2]]
[1]•"th"•""
```

```
>•strapply(text,•"([a-z]*)i([a-z]*)",•c,•ignore.case=T,•perl=T,•backref=-1)¶
[[1]]
[1]•"This"••"Th"••••"is"••••""••••••"first"•"f"
[[2]]
[1]•"this"•"th"•••"is"•••"
```

The function strapply has a variety of other uses, but you will most likely only come to appreciate these when you begin to write your own functions. Thus, I will not go into further details here, but Chapter 4 and the case studies in Chapter 6 will return to strapply whenever appropriate, i.e. repeatedly.

You should now do "Exercise box 3.7: A few more regular expressions" . . .

3.7.6 Merging and Splitting (Vectors of) Character Strings with Regular Expressions

We have seen above in Section 3.7.3 that we can split up character strings using strsplit. While the above treatment of strsplit did not involve regular expressions, this function can handle regular expressions in just about the way that grep and the other functions can. Let us exemplify this by means of a character string that may be found like this in the BNC: every word is preceded by a part-of-speech tag:

```
>•tagtext<-"<w•DPS>my•<w•NN1>mum•<w•CJC>and•<w•DPS>my•<w•NN1>aunt•
  <w•VVD>went•<w•PRP>into•<w•NN1>service"¶
```

If you want to break up this character string such that every word (without the tag) is one element of an overall vector, this is what you could do:

```
>•unlist(strsplit(tagtext,•"<w•...>",•perl=T))¶
[1]•""•••••••••"my•"•••••"mum•"••••"and•"••••"my•"•••••"aunt•"•••
  "went•"•••"into•"•••"service"
```

As you can see, the function looks for occurrences of "<w·" followed by three characters and a ">" and splits the string up at these points. Recall again that unlist is used here to let R return a vector, not a list. While this works fine here, however, life is usually not that simple. For one thing, not all tags might correspond to the pattern "<w· ...>", and in fact not all do. Punctuation marks, for instance, look different:

```
>•tagtext<-"<w•DPS>my•<w•NN1>mum•<c•PUN>,•<w•DPS>my•<w•NN1>aunt•
  <w•VVD>went•<w•PRP>into•<w•NN1>service"¶
```

However, you should be able to find out quickly what to do. This is a solution that would work if the only tag beginnings were "<w·" and "<c·":

```
>•unlist(strsplit(tagtext,•"<[wc]•...>",•perl=T))¶
```

Again, however, life is more complex than this: recall that we have to be able to accommodate portmanteau tags.

```
>•tagtext<-"<w•DPS>my•<w•NN1>mum•<c•PUN>,•<w•DPS>my•<w•AJ0-VVN>beloved•
  <w•NN1>aunt•<w•VVD>went•<w•PRP>into•<w•NN1>service"¶
```

Among the ways of handling this, this is one with alternatives, but without ranges:

```
>•unlist(strsplit(tagtext,•"<[wc]•(...|...-...)>",•perl=T))¶
```

As you can see, the expressions look either for any three characters or for any three characters followed by a hyphen and another three characters. The other possibility involves an expression with a range such that we are either looking for three to seven (three + "-" + three) characters.

```
>•unlist(strsplit(tagtext,•"<[wc]•.{3,7}>",•perl=T))¶
[1]•""•••••••••"my•"••••••"mum•"•••••","•"••••••••"my•"••••••
  "beloved•"•"aunt•"••••"went•"••••"into•"••••"service"
```

One final thing is less nice. As you can see, strsplit splits up at the tags, but it does not remove the spaces after the words. This may not seem like a great nuisance to you but, before you go on reading, look at the last word and think about how this may be problematic for further corpus-linguistic application.

THINK

BREAK

If you could not think of something, consider what would happen if the vector tagtext looked like this:

```
>•tagtext<-"<w•DPS>my•<w•NN1>service•<w•DPS>my•<w•AJ0-VVN>
  beloved•<w•NN1>aunt•<w•VVD>went•<w•PRP>into•<w•NN1>service"¶
```

THINK

BREAK

The problem arises when you apply the above regular expression to this version of tagtext.

```
>•unlist(strsplit(tagtext,•"<[wc]•.{3,7}>",•perl=T))¶
[1]•""•••••••••"my•"•••••"service•"•"my•"•••••"beloved•"•"aunt•"••••
  "went•"••••"into•"••••"service"
```

As you can see, there is one occurrence of "service" with a following space—the first one—and one without a following space—the second one. The second one is not followed by a space because its characters were the last ones in the vector tagtext. The problem is that if you now wanted to generate a frequency list using table, which you already know, R would of course treat "service" and "service·" differently, which is presumably not what anyone would want:

```
>•table(unlist(strsplit(tagtext,•"<[wc]•.{3,7}>",•perl=T)))¶
••••••aunt•••beloved•••into•••my•••service•••service••••went
  ••1••••••1•••••••••1••••••1••••2••••••••••••1•••••••••••1••••••1
```

Thus, what would a better solution look like, i.e. a solution that deletes the spaces?

THINK

BREAK

This is what you could do:

```
>•unlist(strsplit(tagtext,•"•*<[wc]•.{3,7}>•*",•perl=T))¶
[1]•""•••••••••"my"••••••"service"•"my"••••••"beloved"•"aunt"••••
  "went"••••"into"••••"service"
```

Let us end this section by mentioning another corpus-linguistically interesting application of strapply: you can use it to split up character vectors in a way that is different from strsplit. Recall, strsplit required you to provide the characters at which the character string is to be split up, e.g. whitespace. With strapply, you can now split up character vectors, but you do not provide a separator character—you provide the content that is to be retained. Thus, for example, splitting up the vector tagtext into tag-word pairs—not just words!—is a little tricky with strsplit because strsplit will always replace the matched characters. With strapply, this is unproblematic:

```
>•unlist(strapply(tagtext,•"<[^<]*",•perl=T))¶
[1]•"<w•DPS>my•"••••••••••"<w•NN1>service•"
[3]•"<w•DPS>my•"••••••••••"<w•AJ0-VVN>beloved•"
[5]•"<w•NN1>aunt•"•••••••••"<w•VVD>went•"
[7]•"<w•PRP>into•"•••••••••"<w•NN1>service"
```

This line will match every occurrence of a "<" followed by as many characters as necessary but as few as possible to reach the position before the next "<", and these occurrences are returned.

For further study/exploration:

- on how to do approximate, i.e. non-exact, pattern matching: ?agrep¶
- You may wish to explore how strsplit can be used to approximate some of the functionality of the File Utilities in WordSmith Tools 5 (such as splitting up files)
- Spector (2008: Chapter 7)
- on regular expressions in general: Stubblebine (2003), Forta (2004), Good, N.A. (2005), Watt (2005), and the "regular expression bible", Friedl (2006)

3.8 File and Directory Operations

The final section in this chapter introduces a few functions to handle files and folders; some of these will do things you usually perform using parts of your operating system such as Explorer in MS Windows.

One of the most important functions you have already encountered. On Windows, the function choose.files () allows you to interactively choose one or more files for either reading or writing.[16] Other important functions have to do with R's working directory, i.e. the directory that is accessed if no other directory is provided in a given function. By default, i.e. if you haven't changed R's default settings, then this is either R's installation directory or the logged-in user's main data directory. You can easily find out what the current working directory is by using getwd without any arguments at the prompt:

```
>•getwd()¶
[1]•"D:/"
```

If you want to change the working directory, you use setwd and provide as the only argument a character string with the new working directory. Note that the directory must exist before you can set it as the new working directory, setwd does not create a directory for you:

```
>•setwd("C:/_qclwr")•#•if•you•don't•get•a•response,•consider•it•done¶
```

If you now want to find out which files are in your working directory, you can use dir. The default settings of dir are these:

```
dir(path=".",•pattern=NULL,•all.files=F,•full.names=F,•recursive=F)
```

The first one specifies the directory whose contents are to be listed; if no argument is provided, the current working directory is used. The second argument is an optional regular expression with which you can restrict the output. If, for example, you have all 4,054 files of the BNC World Edition in one directory, but you only want those file names starting with "D", this is how you could proceed:[17]

```
>•dir("C:/_qclwr/BNCwe",•pattern="^D")¶
```

The default setting of the third argument returns only visible files; if you change it to TRUE, you would also get to see all hidden files. The default setting of the fourth argument only outputs the names of the files and folder without the path of the directory you are searching. If you set this argument to TRUE, you will get the complete path for every file and folder. By the way, the individual function basename does exactly the opposite: it cuts off all the path information.

```
>•basename("C:/_qclwr/test.txt")¶
[1]•"test.txt"
```

Finally, the default setting recursive=F only returns the content of the current directory. If you set this to TRUE, the contents of all subdirectories will also be returned. This is of course useful if your corpus files come in a particular directory structure. For example, you can use the following line in a script to prompt the user to input a directory and then store the names of all files in this directory and all its subdirectories into one vector called corpus.files.

```
corpus.files<-dir(scan(what="char"),•recursive=T)¶
```

Or, for Windows users even:

```
corpus.files<-dir(choose.dir(),•recursive=T)¶
```

Apart from the number of files in a directory and their names, however, you can also retrieve much more useful information using R. The function file.info takes as input a vector of file names and outputs whether something is a file or a directory, the sizes of the files, the file permissions, and dates/times of file modification, creation, and last access.[18] Thus, the easiest and most comprehensive way to take a detailed look at your working directory is this:

```
>•file.info(dir())¶
```

There are some other file management functions, which are all rather self-explanatory; I will only provide some of the options here and advise you to explore the documentation for more details:

- file.create(x) creates the file(s) in the character vector x;
- dir.create(x) creates the directory that is provided as a character string either as a directory in your working directory or as a relative or rooted directory; thus, dir.create("temp") will create a directory called "temp" in your current working directory; dir.create("../temp") will create a directory called "temp" one directory above your current working directory, and dir.create("C:/_qclwr") will create just that;
- unlink(x) deletes the files/directories in the character vector x; file.remove can only delete files;
- file.exists(x) checks whether the file(s) in the character vector x exist(s) or not;
- download.file(x,•file.choose()) downloads the website at the URL x (i.e. <http://...>) into a user-defined destination file (of course, you can also immediately specify a path).

Sometimes, you need to use a data structure, such as a data frame or a list, again and again and, thus, want to save it into a file. While you can always save data frames as tab-delimited text files, which are easy to use with other software, sometimes your files may be too large to be opened with other software (in particular spreadsheet software such as Microsoft Excel or OpenOffice.org Calc). The option of using raw text files may also not be easily available with lists. Finally, you may want to save disk space and not only save your data, but also compress them at the same time. In such cases, you may want to use save. In its simplest syntax, which is the only one we will deal with here (cf. the documentation for more detailed coverage), you can save any object into a binary compressed file, which will often be much smaller than the corresponding text file and which, if you gave it the extension ".RData", makes your data available to a new R session upon a simple double-click. As an example, consider a data frame aa with the following structure (details such as factor levels, etc. are omitted):

```
'data.frame':•••57437•obs.••of••12•variables:
•$•PERSON•••:•Factor•w/•12•levels•...
•$•PERSON2••:•Factor•w/•2•levels•...
•$•FILE_ID••:•num•...
```

```
•$•FILE_LINE:•int•. . .
•$•VERB•••••:•Factor•w/•6754•levels•. . .
•$•LEMMAS•••:•Factor•w/•2451•levels•. . .
•$•VERBTAG••:•Factor•w/•40•levels•. . .
•$•ASPECT•••:•Factor•w/•2•levels•. . .
•$•ASPECT2••:•Factor•w/•2•levels•. . .
•$•TENSE••••:•Factor•w/•2•levels•. . .
•$•TENSES1••:•Factor•w/•2•levels•. . .
•$•TENSES2••:•Factor•w/•3•levels•. . .
```

This data frame was loaded from a tab-delimited text file with a size of 5.45 MB (5,721,842 bytes). The following function call would save this file into a user-specified file, which should have the extension ".RData" (e.g. <aa.RData>) with a size of 579 KB (593,822 bytes) that, if double-clicked, opens a new instance of R and makes the data available instantly.

```
save(aa,•file=. . .)•#•explore•the•documentation•for•options¶
```

Alternatively, if you have not saved the data structure as an .RData file or only want to open it with starting a new R session, you can just as well use load to access your data again:

```
load(file=. . .)•#•explore•the•documentation•for•options¶
```

You will be prompted to choose a file, and then the data structure(s) will be available. The same can be done with lists that contain all your data, code, and results—recall the advice from Section 3.5 above! Also, you can just save your whole R workspace into a user-definable file with save.image, which is useful to store intermediate results for later processing into an .RData file. This file can either be loaded from within R, or you can just double-click on them to open R and restore your workspace.

Let me finally mention how you would go about and save graphs generated in R. There are two possibilities. The first is to generate the graph and save it from the menu "*File: Save as . . .*" (I usually use the *.png format). The second is to first open a graphics device for the desired output format, plot the file into that graphics device, and then close that device. For example:

```
>•png(filename="03-8_graph.png",•width=600,•height=600)•#•don't•forget•
  the•extension!¶
>••••plot(1:10,•1:10,•type="b")¶
>•dev.off()¶
```

The first line allows you to plot whatever you want to plot into a .png file with the size indicated; when R prompts you for the file name, don't forget the extension ".png" after the file name. The second line plots a simple graph just as an example, and the last line shuts off the graphics device. You can now open the file and look at it.

For further study/exploration:

- on how to read SPSS and Excel files more easily: `?read.spss`¶ and `?read.xls`¶
- on how to read tab-delimited text files: `?read.csv`¶
- on how to produce text representations of objects into files: `?dump`¶
- on how to divert output to a file instead of the screen/console: `?sink`¶
- on how to count the number of fields in lines of a file: `?count.fields`¶
- on how to handle compressed files: `?bzfile`¶ and `?gzfile`¶
- on how to change character encodings of vectors: `?iconv`¶ and `?iconvlist`¶

In the next chapter, we will finally get down to business and apply all these functions to do corpus linguistics.

4

Using R in Corpus Linguistics

Now that corpus linguistics is turning more and more into an integral part of mainstream linguistics, [. . .] we have to face the challenge of complementing the amazing efforts in the compilation, annotation and analysis of corpora with a powerful statistical toolkit and of making intelligent use of the quantitative methods that are available.

(Mukherjee 2007: 141)

In this chapter, you will learn how the functions introduced in Chapter 3 can be applied to the central tasks of a corpus linguist, namely to retrieve and process linguistic data to generate frequency lists, concordances, and collocate displays. Unfortunately, the number of different formats of corpora as well as the number of different tasks you might wish to perform is actually so large that we will not be able to look at all possible combinations. Rather, I will exemplify how to perform different tasks on the basis what are probably the most widely distributed (kinds of) corpora. Let me also mention that there will be only fairly few exercise boxes in this chapter because the case study assignments of Chapter 6 will provide much exercise material on the basis of linguistic data. However, you will still find a variety of think breaks. Finally, an organizational comment: the scripts in the code files provide lines for Mac/Linux users as well as for Windows users. If you work at a Mac/Linux PC, you may want to define the working directory in such a way that you don't have to type much to get to the input and output directories (cf. `setwd` in Section 3.8). If you work at a Windows PC, the code lines will suggest output file names to you that have the same names as the files I provide as a reference. This is so that you can immediately see which file you should compare your own output to. Ideally, you will just add an underscore to the suggested file name and then compare your result. If you ever accidentally overwrite one of the reference files, you can always go back to the zip files you downloaded or even to the website. Now . . . off we go.

4.1 Frequency Lists

4.1.1 A Frequency List of an Unannotated Corpus

Before we begin, let us first make sure that you have nothing left in memory that may influence the result—either you should start a new instance of R or you should clear the memory:

```
>•rm(list=ls(all=T))¶
```

(This is actually something you should do from now on at the beginning of each new section.)

Let us begin with the simplest case. You have a single unannotated corpus file for which you would like to generate a frequency list. We will use the file <C:/_qclwr/_inputfiles/corp_gpl_short .txt>, a few lines from the text of the GNU public license (cf. Figure 4.1):

```
The·licenses·for·most·software·are·designed·to·take·away·your¶
freedom·to·share·and·change·it.·By·contrast,·the·GNU·General·Public¶
License·is·intended·to·guarantee·your·freedom·to·share·and·change·free¶
software--to·make·sure·the·software·is·free·for·all·its·users.·This¶
General·Public·License·applies·to·most·of·the·Free·Software¶
Foundation's·software·and·to·any·other·program·whose·authors·commit·to¶
using·it.·(Some·other·Free·Software·Foundation·software·is·covered·by¶
the·GNU·Library·General·Public·License·instead.)·You·can·apply·it·to¶
your·programs,·too.¶
```

Figure 4.1 The contents of <C:/_qclwr/_inputfiles/corp_gpl_short.txt>.

Since we know that `table` generates a frequency list when applied to a vector, we will proceed as follows. We will load the file and split it up such that all words end up in one vector, to which we then apply `table`. Let us first load the above file; note how the arguments to `scan` aim at avoiding problems with quotes and crosshatches:

```
>•textfile<-scan(choose.files(),•what="char",•sep="\n",•quote="",•
  comment.char="")¶
Read•9•items
```

Let us first look at the beginning or the end of the data structure to see whether the loading has worked as desired:

```
>•head(textfile)¶
[1]•"The•licenses•for•most•software•are•designed•to•take•away•your"
[2]•"freedom•to•share•and•change•it.•By•contrast,•the•GNU•General•Public"
```

```
[3] •"License•is•intended•to•guarantee•your•freedom•to•share•and•change•free"
[4] •"software--to•make•sure•the•software•is•free•for•all•its•users.•This"
[5] •"General•Public•License•applies•to•most•of•the•Free•Software"
[6] •"Foundation's•software•and•to•any•other•program•whose•authors•commit•to"
```

This looks good. If you would like to adopt a very simple definition of what a word is, then you can now generate a frequency list very quickly. Let us assume your frequency list should not be case-sensitive. Therefore, you first change all characters to lower case:

```
>•textfile<-tolower(textfile)¶
```

In a second step, you split up the file at every occurrence of a non-word character (hence the "\\W"):

```
>•words.list<-strsplit(textfile,•"\\W")¶
```

(You may wonder why we didn't use "\\W" as a separator character when we loaded the file with scan. Since scan does not allow multi-byte separators, this is unfortunately not possible.) Since strsplit returns a list but vectors are easier to output, as a third step, you change the list of words into a vector:

```
>•words.vector<-unlist(words.list)¶
```

As a last step, you just generate a frequency list of this vector and sort it in descending order of frequency to get the most frequent words at the top of the file:

```
>•freq.list<-table(words.vector)¶
>•sorted.freq.list<-sort(freq.list,•decreasing=T)¶
```

The table that we obtain from this looks like this:

```
>•head(sorted.freq.list)¶
••••••••••••••••to•software••••••the•••••free••••••and
••••••••9••••••••9••••••••7••••••••5••••••••4••••••••3
```

That is, the values of sorted.freq.list are the frequencies, and their names are the words. This, however, means that if you printed sorted.freq.list into a file, only the frequencies would be saved, not the names. It is easy to fix that, though: we can simply use paste to combine each word and its frequency. Note how this is done here: we define a separator argument, which is the character that intervenes between each word and its frequency, but we do not define a `collapse` argument (other than the default value of NULL) in order to avoid that all word-frequency pairs are glued together:

```
>•sorted.table<-paste(names(sorted.freq.list),•sorted.freq.list,•sep="\t")¶
```

Then, since `sorted.table` is a vector, we output it into a user-defined file using `cat`. We provide a header row and use newlines as separators so that every word-frequency pair gets its own row:

```
>•cat("WORD\tFREQ",•sorted.table,•file=choose.files(),•sep="\n")¶
```

Done. You can now compare this file to the file I provide to check your results, <C:/_qclwr/_out putfiles/04-1-1_output-a.txt>. An alternative way of outputting the results would be this, the result of which you will find in <C:/_qclwr/_outputfiles/04-1-1_output-b.txt>:

```
>•write.table(sorted.freq.list,•choose.files(),•sep="\t",•eol="\n",•
   row.names=names(sorted.freq.list),•quote=F)¶
```

Actually, we could omit the specification of the end of line character since this is the default setting of `write.table` anyway, and we will do so in the future.

Let us now turn to a few useful improvements. First, you may sometimes not be interested in obtaining *all* frequencies. For example, you may want to introduce a minimum threshold value *m* to only list word forms that occur at least *m* times. How, for example, would you exclude *hapax legomena*—words that occur in your corpus just once—from your frequency list?

THINK

BREAK

The answer to this one is simple. You can exclude *hapax legomena* with the following line of code that you use before you output the table into a file:

```
>•sorted.freq.list<-sorted.freq.list[sorted.freq.list>1]¶
```

Second, if you look at the frequency list, there is one other aspect that is not so nice: we have nine empty elements in there. You can easily see that they result from the generation of the list `words .list` at cases where two non-word characters are directly next to each other. Now, try to find out how can you get rid of these empty elements; you need to add only one character to the code we have written so far.

THINK

BREAK

The answer is probably very obvious: since we want to handle cases where there is one non-word character *or more*, the only thing we need to change is the way we split up the corpus files:

```
>•words.list<-strsplit(textfile,•"\\W+)•#•the•change•is•the•"+"¶
```

If you now proceed as before, you will see that we now get the desired output. Finally, in Section 2.2, I mentioned that very often frequency lists are reduced on the basis of so-called stop lists in order to exclude frequent but irrelevant function words. Let us assume you have the following stop list:

```
>•stop.list<-c("the",•"of",•"and",•"it",•"is",•"be",•"are")¶
```

How do you now exclude the frequencies of these words from your frequency list?

THINK

BREAK

This may have been difficult, but if you look at it, you will understand it immediately:

```
>•words.vector.2<-words.vector[!(words.vector•%in%•stop.list)]¶
```

The expression words.vector•%in%•stop.list determines for each word of words.vector whether it is in stop.list.

```
>•head(words.vector•%in%•stop.list)•#•just•for•the•first•six•words¶
[1]••TRUE•FALSE•FALSE•FALSE•FALSE••TRUE
```

But then we cannot use this with subsetting because we don't want the set of words that occur in stop.list—we want the opposite. So we change it into the opposite by means of the negation operator we know from our logical expressions, the exclamation mark. This can now be used to subset words.vector as shown above, and then you can proceed as before.

```
>•head(!(words.vector•%in%•stop.list))•#•just•for•the•first•six•words¶
[1]•FALSE••TRUE••TRUE••TRUE••TRUE•FALSE
```

Finally, another nice solution makes use of strapply again (from the package gsubfn). Once you have loaded the file contents into textfile as above, why not just write this to get an alphabetically sorted table?

```
>•library(gsubfn)¶
>•table(unlist(strapply(textfile,•"\\w+",•perl=T)))¶
```

Since you will usually want to process such data further and since a spreadsheet program would be the usual way of doing so, let us briefly look at how you would import this frequency list into a spreadsheet program, OpenOffice.org 3.0 Calc. First, you open OpenOffice.org Calc. Second, you click on "*File: Open*". Third, you navigate to the directory in which the frequency list file is located, which should be <C:/_qclwr/_outputfiles>. Fourth, in the dropdown field "*Files of type*", you choose "*Text CSV (*.csv; *.txt;*.xls)*", pick the file from the list of files above, and click "*Open*". Fifth, you will then see the text import wizard. Note in particular that:

- you should delete the text delimiter;
- you should use tab stops as the only separators (not spaces, though that may be needed on other occasions);
- you do *not* tick the box "Merge delimiters" so that two tab stops next to each other will be treated as two separators, not one.

Click on "OK" and you are done.

4.1.2 A Reverse Frequency List of an Unannotated Corpus

As a second example, let us assume you would like to generate a case-insensitive reverse frequency list of the same file as in Section 4.1.1. We can proceed just about as before but now we must reverse the characters of each word before we generate the table. Again, you first load the file, change its content all into lower case, split it up on the basis of non-word characters, and change the list into a vector:

```
>•textfile<-tolower(scan(choose.files(),•what="char",•sep="\n",•quote="",•
  comment.char=""))¶
Read•9•items
>•words.list<-strsplit(textfile,•"\\W+")¶
>•words.vector<-unlist(words.list)¶
```

Then, you use strsplit to split up each word into its individual letters; not specifying a character at which the strings are to be split up makes R split them up character by character:

```
>•list.of.letters<-strsplit(words.vector,•"")¶
>•list.of.letters[1:3]¶
[[1]]
[1]•"t"•"h"•"e"
[[2]]
[1]•"l"•"i"•"c"•"e"•"n"•"s"•"e"•"s"
[[3]]
[1]•"f"•"o"•"r"
```

As a next step, we reverse the elements of every vector in the list. There are two ways of doing this; one is less elegant (in R at least), the other is more elegant. The less elegant way involves accessing each component of the list, which is a vector with the letters of one word, with a for-loop. First, we create a vector for the results of the reversing process, which of course needs to have as many elements as there are words to reverse:

```
>•words.reversed<-vector(length=length(list.of.letters))¶
```

Then, we access each component of the list, reverse the letters of each word using rev, store them in the vector reserved for the results, and iterate:

```
>•for•(i•in•1:length(list.of.letters))•{¶
+••••current.vector.of.letters<-list.of.letters[[i]]¶
+••••rev.current.vector.of.letters<-rev(current.vector.of.letters)¶
+••••words.reversed[i]<-paste(rev.current.vector.of.letters,•collapse="")¶
+•}¶
```

Finally, we again generate a frequency list of this vector and, if we do not want to sort it first, just save it:

```
>•rev.freq.list<-table(words.reversed)¶
>•rev.freq.table<-paste(names(rev.freq.list),•rev.freq.list,•sep="\t")¶
>•cat(rev.freq.table,•file=choose.files(),•sep="\n")¶
```

The more elegant version does away with the for-loop and uses functions from the apply family. First, we use sapply to apply rev to every element of the list:

```
>•rev.list.of.letters<-sapply(list.of.letters,•rev)¶
>•rev.list.of.letters[1:3]¶
[[1]]
[1]•"e"•"h"•"t"
[[2]]
[1]•"s"•"e"•"s"•"n"•"e"•"c"•"i"•"l"
[[3]]
[1]•"r"•"o"•"f"
```

Then, we merge the elements of each vector in the list by applying paste to each vector in the list and forcing R to return a vector with sapply; this we can tabulate before saving the result, which you can compare to the output file in <C:/_qclwr/_outputfiles/>. Note one important aspect of sapply here: it takes an argument that is actually an argument of paste, namely collapse="":

```
>•words.reversed<-sapply(rev.list.of.letters,•paste,•collapse="")¶
>•rev.freq.list<-table(words.reversed)¶
>•rev.freq.table<-paste(names(rev.freq.list),•rev.freq.list,•sep="\t")¶
>•cat(rev.freq.table,•file=choose.files(),•sep="\n")¶
```

While both these approaches work well, you might object to both on the grounds that they are not particularly efficient. Can you see why?

THINK

BREAK

As you may have found out yourself, both above approaches are not optimal because they involve reversing of all word *tokens* although it would actually only be necessary to apply it to all word *types* in the table. Thus, the better solution would actually be to first generate a frequency table as in Section 4.1.1, and then only reverse the names of the frequencies. This is how you could do it all:

```
>•textfile<-tolower(scan(choose.files(),•what="char",•sep="\n",•quote="",•
  comment.char=""))¶
>•words.vector<-unlist(strsplit(textfile,•"\\W+"))•#•split•up•into•words¶
>•freq.list<-table(words.vector)•#•generate•and•sort•a•table¶
>•word.types<-names(freq.list)¶
>•list.of.letters<-strsplit(word.types,•"")•#•as•before¶
>•rev.list.of.letters<-sapply(list.of.letters,•rev)•#•as•before¶
>•words.types.reversed<-sapply(rev.list.of.letters,•paste,•
  collapse="")•#•as•before¶
>•names(freq.list)<-words.types.reversed¶
```

You would now just have to output that as above. Again, the different procedure does not make much of a difference here, but with larger corpora it is useful to always use less computationally demanding approaches.

4.1.3 A Frequency List of an Annotated Corpus

Let us now turn to generating a frequency list of the words in a corpus file that has some annotation, a file with SGML annotation of the kind you find in the BNC World Edition. Again, the overall approach will be as before, but this time we need to get rid of the annotation before we tabulate. First, we load the file <C:/_qclwr/_inputfiles/corp_bnc_sgml_4.txt>; note that here we don't need to worry about quotes, because the sep argument ("\n") makes sure that quote is set to "".

```
>•textfile<-scan(choose.files(),•what="char",•sep="\n")¶
Read•2873•items
```

Let us assume we wish to do a case-sensitive frequency list; thus, we do not change anything to lower (or upper) case. Second, since we only want to include the actual corpus sentences (rather than also the headers, etc., a bug which distorted MonoConc Pro's results!), we next reduce the corpus file to only the corpus sentences. In the BNC, the sentences are numbered as follows: <s·n="1">, <s·n="2">, . . . Thus, we can use this annotation to discard all the lines in the file that do not have this annotation:

```
>•tagged.corpus.sentences<-grep("<s·n=",•textfile,•value=T)¶
```

Third, we split up the file at every occurrence of a tag, which all start with "<" and end with ">". This will utilize the information that the tags provide—for example, it will preserve the information that *apart from* is tagged as one word rather than two but at the same time take out all the tags:

```
>•words.vector<-unlist(strsplit(tagged.corpus.sentences,•"<.*?>",•perl=T))¶
>•head(words.vector)¶
[1]•""•••••""•••••"the•"•"bank"•"."•••••""
```

We still need to clean up, however: we should get rid of the empty character strings and we should delete any leading/trailing spaces. As to the former, we do this in a way similar to Exercise box 3.6; as to the latter, we use nchar to retain only strings with more than 0 characters:

```
>•words.vector<-gsub("(^•+|•+$)",•"",•words.vector,•perl=T)¶
>•words.vector<-words.vector[nchar(words.vector)>0]¶
```

Can you think of an alternative to the second line using grep?

THINK

BREAK

We would just have to look for any word character:

```
>•words.vector<-grep("\\w",•words.vector,•perl=T,•value=T)¶
```

We can now tabulate and save in the way discussed in Section 4.1.1.

Another version, which is shorter to program but takes longer to execute, uses strapply from the package gsubfn. It simply retrieves from the tagged sentences all character sequences that follow the closing angular bracket of a word tag and do not contain an opening angular bracket (which would be the beginning of the next tag). Then you only need to take care of trailing spaces, and then you can again use table to generate the frequency list and proceed as above in Section 4.1.1. The results will differ slightly from the above approach in the sense that this approach will not include punctuation marks, but the word frequencies should be the same:

```
>•words.vector<-unlist(strapply(tagged.corpus.sentences,•">([^<]*)",•perl=T))¶
>•words.vector<-gsub("•$",•"",•words.vector,•perl=T)¶
```

4.1.4 A Frequency List of Tag-word Sequences from an Annotated Corpus

The next example concerning frequency lists will be a little bit more complex. We will again generate a case-insensitive frequency list, but this time we will use a for-loop to read in more than one corpus file. Also, we will delimit the range of data to be included (by again excluding headers, etc.). We will first combine all sentences from all corpus files into one vector, and then use strsplit to arrive at a vector of tag-word sequences to which we can apply table.

First, we choose the files that are to be included in the frequency list; for now we will use the two files <corp_bnc_sgml_2.txt> and <corp_bnc_sgml_4.txt> in the directory <C:/_qclwr/ _inputfiles>:

```
>•corpus.files<-choose.files()¶
```

Then, we use a for-loop to first access every file in our vector corpus.files. However, since, this time, we want to generate a frequency list of several corpus files, we also need to reserve a data structure for the whole corpus, i.e. for the character vector containing the result of merging all the different corpus files into one corpus.

```
>•whole.corpus<-vector()¶
```

This is an important step you have to remember well. Nearly always when you want to look at more than one file, you go through the files with a loop, but must reserve data structures for what is going to be stored in the loop. Another rule of thumb: if you need to store sequences of elements, but don't have to keep these correlated with something else, you will mostly reserve vectors; however, sometimes when you need to do more complex things, lists can be more useful. The most straightforward possibility to generate the frequency list is the following: We load each corpus file into a vector called current.corpus.file, using the command scan with newlines as separators; the argument quiet=T suppresses the output of the number of elements read. Since we do not want to distinguish between small and capital letters, we also change all corpus lines into lower case:

```
>•for•(i•in•corpus.files)•{¶
+••••current.corpus.file<-scan(i,•what="char",•sep="\n",•quiet=T)¶
+••••current.corpus.file.2<-tolower(current.corpus.file)¶
```

Note how this loop is different from before. Last time—in Section 4.1.2—we accessed each element by establishing a counter i (taking on values from 1 to the maximum number) and then used the counter to access every element by subsetting. If we had applied this logic here, this is what we would have done:

```
for•(i•in•1:length(corpus.files))•{¶
•••current.corpus.file<-scan(corpus.files[i],•what="char",•sep="\n",•quiet=T)¶
```

Why does the first solution also work and what is the difference to the earlier one? The answer to the first question is that a for-loop was defined above as for•(var•in•seq)•{, and a vector, such as corpus.files, can be understood as a sequence. The answer to the second question is, thus, we do not use subsetting to access the individual values of the vector—we access the values directly instead. What is the benefit of each approach? The former approach is called for if you still need the counter that goes from 1 to ... for something else such as, for example, storing a result at a particular position in a vector that is defined by the counter. The present approach is more appealing—given its simpler code—if the counter is not needed any more. Here we just want to combine lines without the need to do anything else with the counter, and we can choose the shorter version; the code that follows will be based on that decision.

Then, we output the name of the corpus files currently processed. The latter will slow down the whole process a tiny little bit, but allows us to monitor progress and, in the case of problems, see where problems occurred:

```
+••••cat(i,•"\n")•#•output•a•'progress•report'¶
```

In the next step, we again must tell R that we do not wish to include the header, utterance tags, etc. in our counts. Thus, we again reduce the current corpus file to only those lines with a BNC-style sentence number ("<s·n="1">", "<s·n="2">", etc.).[1] But of course, we do not want to use the sentence number in our counts, which is why after we used sentence numbers to discard the header, we delete them:

```
+••••current.sentences<-grep("<s•n=",•current.corpus.file.2,•perl=T,•
  value=T)¶
+••••current.sentences<-sub("<s•n=.*?>",•"",•current.sentences,•perl=T)¶
```

The results of these operations are then stored in / appended to the vector whole.corpus and the loop is completed:

```
+••••whole.corpus<-append(whole.corpus,•current.sentences)¶
+•}¶
```

This vector now contains all the sentences in the corpus files. In the next step, we must access the individual words and their tags of each sentence. Since, however, we again have tag information with non-word characters such as "<", ">", "-", "=", etc., we cannot use strsplit with "\\W+". We can, however, make use of the fact that every tag-word sequence in the BNC starts with "<" and, thus, write:

```
>•all.words.1<-unlist(strsplit(whole.corpus,•"<"))¶
```

This is what we have at the moment:

```
>•all.words.1[1:5]¶
[1]•""•••••••••••"w•at0>the•"•"w•nn1>bank"•"c•pun>."••••""
```

As we can see, there is still room for improvement: we can disregard all elements that do not begin with "w·" because they are tags of punctuation marks (starting with "<c·") or other forms of annotation (such as "<pause>", "<trunc>", etc.). In addition, we can also delete all "w·" at the beginnings of our word-tag sequences:

```
>•all.words.2<-grep("^w•",•all.words.1,•perl=T,•value=T)¶
>•all.words.3<-gsub("^w•",•"",•all.words.2,•perl=T)¶
```

Finally, we should take care of the fact that words are followed by a space when a different word follows but are not when a punctuation mark follows. Thus, our procedure so far would treat the string "table" differently depending on whether it was used in the corpus file in "This table is new" and "This is a new table." because the former would have resulted in "table·" whereas the latter would have resulted in "table". Thus, let us delete string-final spaces:

```
>•all.words.4<-gsub("•$",•"",•all.words.3,•perl=T)¶
```

Of course, this could also have been addressed in the line in which generated all.words.1 by splitting up not at "<" but at "·*<". Anyway, we're done:

```
>•sorted.freq.list<-sort(table(all.words.4),•decreasing=T)¶
>•head(sorted.freq.list)¶
all.words.4
at0>the••vbz>is••pnp>it•cjc>and•••at0>a••prf>of
•••4975••••3000••••2850••••2501••••2119••••2031
>•write.table(sorted.freq.list,•choose.files(),•sep="\t",•quote=F)•#•compare•
  with•output•file•<04-1-4_output-a.txt>¶
```

There are a few possibilities to prettify the output. First, you may want to further process the output in a spreadsheet software, which is why separating tags and words by tab stops is preferable. While you could do this easily in OpenOffice.org Calc, it is just as easy in R by again using names. If you then save the frequency list, it will have the desired format (cf. <C:/_qclwr/_outputfiles/04-1-4_output-b.txt>):

```
>•names(sorted.freq.list)<-sub(">",•"\t",•names(sorted.freq.list),•perl=T)¶
```

Alternatively, you may be interested in sorting the output alphabetically so that the different tag frequencies of one word form are listed together. However, there are two problems. On the one hand, as you know, if you apply `sort` to the data structure generated by `table`, R sorts the *values* of the table, not the *names of the values* in the table, which are of course the words. On the other hand, even if you could apply `sort` to the names, you would still not get the word forms together because the tags precede the word forms. Before you read on, think about how you could solve this problem.

THINK

BREAK

A shorter way of doing this, which uses fewer lines of code and fewer data structures, would be the following; make sure you understand how this works (cf. <C:/_qclwr/_outputfiles/04-1-4_output-c.txt>) and note how many different tags some word have (e.g. *around* and *as* have five each):

```
>•sorted.freq.list<-sort(table(all.words.4),•decreasing=T)¶
>•names(sorted.freq.list)<-sub("([^>].*?)>(.*$)",•"\\2>\\1",•
  names(sorted.freq.list),•perl=T)¶
>•sorted.freq.list.2<-sorted.freq.list[order(names(sorted.freq.list),•
  decreasing=F)]¶
```

The usual alternative approach with `strapply` could look like this. Immediately after the for-loop with which the character vector `whole.corpus` was filled, you enter this:

```
>•library(gsubfn)¶
>•all.word.tag.sequences<-strapply(whole.corpus,•"<[^<]*",•perl=T)¶
```

By now, you are probably able to understand this without much explanation. The regular expression matches all occurrences of an opening angular bracket and all following characters that are not themselves opening angular brackets and returns these in a list, which has as many character vectors as elements as `whole.corpus` has elements. Now, the only remaining steps before we can begin to generate and customize the frequency list are to take care of unwanted tags and trailing spaces. The first of the two lines below converts `strapply`'s list output into a vector and then discards all elements that do not contain the beginning of a proper POS tag; the second line deletes trailing spaces:

```
>•all.word.tag.sequences<-grep("<w•",•unlist(all.word.tag.sequences),•
  perl=T,•value=T)¶
>•all.word.tag.sequences<-gsub("•$",•"",•all.word.tag.sequences,•perl=T)¶
```

The solution above has one not-so-nice characteristic. It is based on putting all the sentences of the corpus into the computer's working memory. This is unproblematic in the present case, because the corpus is so small, but with the whole BNC, you'd have to commit more than six million sentences into memory (and you could see how much time `strapply` already needed). Alternative approaches could either add all the sentences of each file into one interim file (with cat(. . . , •append=T)) or you could compute a frequency list for each file *within the loop* and then amalgamate the frequency lists (cf. Exercise box 3.1, assignment 10—now you know why we did this!). This way, R would only have to bear in mind the growing frequency list—each type just once with a number—rather than hundreds of thousands of sentences or many tokens. Just a thought . . .

4.1.5 A Frequency List of Word Pairs from an Annotated Corpus

4.1.5.1 Retrieving Word Pairs

While some, though certainly not all, previous examples are possible with ready-made concord-ancers, let us now turn to examples where R excels over such software. This example will not only be a little bit more complex, but also one whose application to larger corpora may require quite a bit of RAM and time (even on well-equipped computers), depending on how it is implemented. In addition, we will play around with a few alternative ways of doing the same thing in order to give you a feel of what options you have and how they differ.

Let us assume we would like to generate a case-insensitive frequency list of pairs of adjacent words in an annotated corpus file such as a part of the tagged Brown corpus. In this corpus, a corpus of written American English in the 1960s, each line starts with a file identifier, a colon, and a line number followed by a space. The rest of the line then consists of the words and punctuation marks of that line, each of which is directly preceded by a space and directly followed by an underscore and a part-of-speech tag. The two lines shown in Figure 4.2 exemplify this annotation:

```
[1]·"|sa01:1·the_at·fulton_np·county_nn·grand_jj·jury_nn·said_vbd·...¶
[2]·"|sa01:2·the_at·jury_nn·further_rbr·said_vbd·in_in·term-end_nn·...¶
```

Figure 4.2 The first two lines of the tagged version of the Brown corpus.[2]

The general approach will be this: we load the file, convert everything to lower case, remove the tags, and do a little cleaning up; most of this is doable using regular expression functions with grep and sub or gsub. Then, there are two conceivable ways of proceeding: we either *retrieve* word pairs, or we *generate* them, and for expository purposes, we will deal with both ways and will cover quite a few issues along the way.

First, we load a file that is analogous to the tagged version of the Brown corpus from <C:/_qclwr/_inputfiles/corp_brown-tagged.txt> and, since we do not need to distinguish small and capital letters, immediately convert the data to lower case:

```
>•corpus.file<-tolower(scan(choose.files(),•what="char",•sep="\n"))¶
Read•10•items
```

Second, we remove the line-initial annotation by deleting everything from the beginning of the line up to the first space. To that end, we first use the anchor "^" for the beginning of the line. After the beginning of the line, there can be any number of characters, which we represent using a wildcard character for every character "." followed by the quantifier "*" and the space we know separates the identifier from the corpus line. However, in order to avoid R matching the longest possible sequence, we use the question mark for ungreedy matching:

```
>•corpus.file.2<-sub("^.*?•",•"",•corpus.file,•perl=T)¶
```

Note that with this we have also deleted the space after the identifier. On other occasions, however, one would actually leave them in because the spaces are the regular separator characters in the corpus and it is always nice to have everything cleanly surrounded by the separator characters we will use later so split things up. On such an occasion, one might even introduce spaces as separator characters at the *end* of the line, which you could do as follows (because one space is the default separator of paste).

```
corpus.file.2<-paste(corpus.file.2,•"")¶
```

We then also define an output file.

```
>•output.file<-choose.files()¶
```

Since this section is about *retrieving* word pairs, this is what we now do (in each line). We cannot use grep on each line here, however, because grep would only return the whole vector element. We also cannot use regexpr because that would only return the first match in each line. But we can use gregexpr. The simplest and best approach might seem to look for "\w+·\w+" . . . But it isn't . . . It is not for several reasons.

First, corpora are never that simple. For example, one file may contain the following line, which we may call test.vector:

```
>•test.vector<-"•henry_np•l._np•bowden_np•was_bedz•not_*•listed_vbn•on_in•
  the_at•petition_nn•as_cs•the_at•mayor's_nn$•attorney_nn•._.•"¶
```

If we apply our expression to this line of the corpus, this is what happens:

```
>•(matches<-gregexpr("\\w+•\\w+",•test.vector,•perl=T))¶
[[1]]
[1]••2•13•27•42•59•78
attr(,"match.length")
[1]•10•13•13•16•18•12
```

As you can see, the first match is not the first pair of what we consider words: it's "henry_np l". Why is this so? This is so because the period after the "l" is not an instance of "\w", it's not a word character. Thus, we must be more comprehensive in our definition of what to match. Recall, it is absolutely imperative that you know what your corpus looks like, what kinds of characters occur and where. If you miss out on one or more of the special characters ("$", "*", etc.), you will distort the results. (Note again that you must not blame that on R: other programs also require you to define separator characters and the like explicitly, but, as you will see in Section 4.5.1 below, R also makes it very easy to find such characters.)

Second, we have seen in Section 3.7.5 above that once a string has been matched and consumed, it is not available for a second match. Thus, even if you improved the above expression to handle the non-word characters, it would fail to match the second word twice: once as the second word of the first word pair, and once as the first word of the second word pair. We need to proceed differently, and you can already guess we will use lookahead (as above), which we know doesn't consume. However, there seems to be one contradiction here: we don't want to consume—because otherwise we can't handle overlapping matches—but we want to capture/consume, because if we don't capture/consume (the second word), how can we print or count it?

We will employ a little trick: we will use lookahead to avoid consuming while at the same time use capturing for back-references. To show you how this works, let me first use a shorter search pattern. Let's assume that we have a vector v that looks just like our lines but has just letters between spaces and that we want to generate all pairs of adjacent letters: "a•b", "b•c", etc. This is what we do:

```
>•v<-"•a•b•c•d•e•f•"¶
>•step1<-gsub("•([a-z])(?=•([a-z]))",•"\\1•\\2\n",•v,•perl=T)¶
>•step2<-gsub("\n.*?$",•"",•step1,•perl=T)¶
>•step3<-unlist(strsplit(step2,•"\n"))¶
```

What does this do and why? In the first step, we look for a space, followed by any letter, and we put the expression matching the letter into parentheses so that we can refer back to it using "\\1". Then, a positive lookahead follows: we only want the expression to match if the back-referenced letter is followed by a space and another letter. The trick is now that, even though this is all in a lookahead, i.e. not consumed and available for a second match, this second letter is also put into parentheses and thus available for back-referencing with "\\2". And then we replace all this with the first back-reference (i.e. the first letter), a space (as a separator), and the second back-reference (i.e. the second letter). The nice thing is that since in the first go only the first letter is consumed—the "a"—the expression can then go on replacing with the "b" being the next first letter of a potential pair. The result of this is step1:

```
>•step1¶
[1]•"a•b\nb•c\nc•d\nd•e\ne•f\n•f•"
```

As you can see, however, the final "f" is never replaced: it simply never matches because it is not followed by a space and another character, but we of course don't want it. So we replace it by replacing the shortest sequence of characters between a line break and the end of the string. This shortest sequence—note the non-greedy matching!—is of course the sequence from the last line break in the string, which is why the above line yields the following string as step2:

```
>•step2¶
[1]•"a•b\nb•c\nc•d\nd•e\ne•f"
```

Finally, as you can see we now have one string only. However, we inserted line breaks so we could either print this into an interim file or, because we have already seen that this is not particularly elegant, use strsplit to get a vector of word pairs, step3:

```
>•step3¶
[1]•"a•b"•"b•c"•"c•d"•"d•e"•"e•f"
```

We should now be ready to apply the same logic to our real case. The only thing we need to do is replace "[a-z]" by what we want to use for recognizing words. We make use of the facts that (i) the underscore is the separator between a word and its tag and (ii) even punctuation marks are separated from the preceding word by a space. A word-tag sequence can therefore be defined as "[^_]+_[^•]+" and we write:

```
>•all.word.pairs.1<-gsub("([^_]+_[^•]+)(?=•([^_]+_[^•]+))",•"\\1•\\2\n",•
   corpus.file.2,•perl=T)¶
>•all.word.pairs.2<-gsub("\n.*?$",•"",•all.word.pairs.1,•perl=T)¶
>•all.word.pairs.3<-unlist(strsplit(all.word.pairs.2,•"\n"))¶
```

If we now check the first line, this looks much better:

```
>•head(all.word.pairs.3)¶
[1]•"since_cs•the_at"•••••••••••••"the_at•psychiatric_jj"
[3]•"psychiatric_jj•interview_nn"•"interview_nn•,_,"
[5]•",_,•like_cs"•••••••••••••••••"like_cs•any_dti"
```

Thus, we can now finally generate our frequency list and save the data as desired:

```
>•freq.table<-table(all.word.pairs.3)¶
>•sorted.freq.table<-sort(freq.table,•decreasing=T)¶
>•write.table(sorted.freq.table,•file=choose.files(),•sep="\t",•quote=F)¶
```

Another nice way, which relies on Gabor Grothendieck's package gsubfn, makes use of strapply. Its logic is about the same as the one outlined above, and while its output still requires some processing, it's a very short and efficient approach:

```
>•library(gsubfn)¶
>•all.word.pairs<-unlist(strapply(corpus.file.2,•"•*([^_]+_[^•]+)
   (?=•([^_]+_[^•]+))",•backref=-2,•
   (?=•([^_]+_[^•]+))",•paste,•perl=T))¶
```

4.1.5.2 *Generating Word Pairs*

In the last session, we tried to get the frequency list of word pairs by *retrieving* word pairs; in this section, we deal with the same problem by going through each line of text and *generating* the pairs of all neighboring words; this should sound to you as if we need a for-loop, but the more difficult question is what to do next. Either we begin the counting of word pairs immediately after the generation (i.e. within the for-loop) or we first save the word pairs somewhere and then do the counting later (i.e. after the for-loop). Since the latter approach is more in line with the policy of making small steps and saving interim results, we will choose the latter option for now. The first lines of code are nearly the same as above, but this time we do delete the first space after the line identifier:

```
>•corpus.file<-tolower(scan(choose.files(),•what="char",•sep="\n"))¶
>•corpus.file.2<-sub("^.*?•",•"",•corpus.file,•perl=T)¶
```

We then split up the lines of the corpus files using strsplit and spaces to get a list with as many elements as there are corpus lines. Each of these elements is a character vector with as many elements as there are words in the line.

```
>•corpus.file.3<-strsplit(corpus.file.2,•"•")¶
>•corpus.file.3[1]•#•look•at•the•first•element¶
[[1]]
•[1]•"since_cs"•••••••••"the_at"•••••••••••"psychiatric_jj"•••"interview_nn"
•[5]•",_,"•••••••••••••"like_cs"••••••••••"any_dti"••••••••••"other_ap"
•[9]•"interview_nn"•••••",_,"••••••••••••"depends_vbz"•••••••"on_in"
[13]•"communication_nn"•",_,"••••••••••••"it_pps"•••••••••••"is_bez"
[17]•"significant_jj"••••"to_to"••••••••••"note_vb"•••••••••"that_cs"
[21]•"the_at"•••••••••••"therapist_nn"•••••"in_in"••••••••••"this_dt"
[25]•"interview_nn"•••••"was_bedz"•••••••••"a_at"•••••••••••"man_nn"
[29]•"of_in"•••••••••••"marked_vbn"•••••••"skill_nn"••••••••"and_cc"
[33]•"long_jj"•••••••••••"experience_nn"••••"._."
```

Then, we define two output files, one to take up all word pairs and one that will contain the real frequency list:

```
>•interim.file<-choose.files()•#•<04-1-5-2a_interim.txt>¶
>•output.file<-choose.files()•#•<04-1-5-2b_output.txt>¶
```

The next step now involves the generation of word pairs. To that end, we need a first for-loop to access each vector of words and punctuation marks of each line (i.e. the list corpus.file.3). Within this loop, we need a second one, which will take each word of the line (from left to right)—let's call it A—and print it together with the next word—let's call it B—into the output file. However, since the last word of a line does not have a next word to print it with, we must make sure that the first loop never accesses the last word of any line. We first start our outer loop and also print out a progress report, because with huge files it helps us to see whether R is still working and how far it got. Since we may have forgotten the number of lines the file has, we print out the percentage.

```
>•for•(i•in•1:length(corpus.file.3))•{•#•go•through•the•list¶
+••••cat(i/length(corpus.file.3),•"\n")•#•progress•report¶
```

Then, we access each line in our list, go through each vector representing a line, and access all words *but the last one*:

```
+••••current.line<-corpus.file.3[[i]]•#•access•each•line¶
+••••for•(j•in•1:(length(current.line)-1))•{•#•go•up•to•the•penultimate•
  word,•thus•"-1"¶
```

Lastly, we print the first word, a space, the second word, and a line break into the output file, and then we complete both loops:

```
+••••••cat(current.line[j],•"•",•current.line[(j+1)],•"\n",•sep="",•
  file=output.file,•append=T)•#•print•into•file¶
+••••}•#•end•of•for:•go•through•the•line¶
+•}•#•end•of•for:•go•through•the•list¶
>•all.word.pairs<-scan(interim.file,•what="char",•sep="\n")•#•load•interim¶
```

Of course, we did not have to print all the word pairs into an interim file within the innermost loop—we could also have generated a vector storing all word pairs and processed that vector *after* both for-loops. In that case, we would have (i) created an empty vector all.word.pairs before entering the first loop and would have (ii) stored each word pair in that vector while we iterate (cf. the code file for the corresponding code). Be that as it may, as a result, we now have an output file with all word pairs per line of the part of the Brown corpus, which we can load and from which we can generate a frequency-list file as before:

```
>•freq.table<-table(all.word.pairs)¶
>•sorted.freq.table<-sort(freq.table,•decreasing=T)¶
>•write.table(sorted.freq.table,•file=output.file,•sep="\t",•quote=F)¶
```

A final comment is in order. The strategy we have adopted above may seem overly complex when compared to how a ready-made concordance program would handle this, or even compared to a different strategy in R. For example, it would of course have been possible to collapse the complete corpus into a single vector rather than a list of vectors and then just go through only a single for-loop to generate all word pairs:

```
>•corpus.file.3b<-unlist(strsplit(corpus.file.2,•"•",•perl=T))¶
>•for•(j•in•1:(length(corpus.file.3b)-1))•{¶
+••••cat(corpus.file.3b•[j],•"•",•corpus.file.3b•[(j+1)],•"\n",•sep="",•
  file=output.file,•append=T)¶
+•}¶
```

Before you read on, consider why this strategy may in fact be less desirable than it seems.

THINK

BREAK

This simpler strategy has the disadvantage that it would result in the formation of word pairs which cross line boundaries, effectively forming pairs of the last word of a line with the first word of the next line. This may be unproblematic or even desirable here, since the line breaks in the Brown corpus are rather arbitrary—in corpora where lines correspond to sentences, however, this may yield results that are undesirable.[3] Note also that this shortcut strategy is how most ready-made programs produce their lists of *n*-grams by default without leaving you a choice (e.g. kfNgram, MLCT concordancer, Collocate 0.5.[4])

4.1.6 A Frequency List of an Annotated Corpus (with One Word Per Line)

Consider as one of the last examples the British component of the International Corpus of English (ICE-GB), in which each line contains some annotation and sometimes a word in curly brackets.[5] The beginnings of sentences or utterances are marked with "PU", for "parse unit"; cf. Figure 4.3 for the contents of the file <C:/_qclwr/_inputfiles/corp_ice-gb.txt>.

It is difficult to see how you would be able to load the file into regular concordance software in such a way that it can be properly processed. But you immediately see that words are between curly brackets, which means tokenization—the identification and separation of the words to be counted—has already been done for us. Thus, you just enter the following to obtain all the words of the file in a single vector that can be easily tabulated with `table` or searched with `grep`.

```
>•current.corpus.file<-scan(choose.files(),•what="char",•sep="\n")•#•load•
  the•corpus•file¶
>•current.corpus.file.2<-grep("}$",•current.corpus.file,•perl=T,•
  value=T)•#•identify•the•lines•which•contain•words•in•curly•brackets¶
>•current.corpus.file.3<-sub(".*{",•"",•current.corpus.file.2,•perl=T)•#•
  delete•everything•before•the•opening•curly•bracket¶
>•current.corpus.file.4<-sub("}$",•"",•current.corpus.file.3,•perl=T)•#•
  delete•the•closing•curly•bracket•at•the•end¶
>•head(current.corpus.file.4)¶
[1]•"Did"
[2]•"we"
[3]•"actually"
[4]•"<unclear-words>"
```

```
[<#117:1:A>·<sent>]
PU,CL(main,inter,intr,past,incomp,supersede)
·INTOP,AUX(do,past)·{Did}
·SU,NP()
··NPHD,PRON(pers)·{we}
·A,AVP(ge)
··AVHD,ADV(ge)·{actually}
[<unclear>]
·INDET,?(ignore)·{<unclear-words>}
[</unclear>·<$?>]
[<#118:1:?>·<sent>]
PU,EMPTY
[<O>·<EXTM(begin)>·inarticulate-noise·</O>·<EXTM(end)>·<$A>]
[<#119:1:A>·<sent>]
PU,NONCL()
·ELE,NP()
··DT,DTP()
···DTPS,NUM(card,sing)·{a·hundred}
··NPHD,N(com,sing)·{problem}
```

Figure 4.3 The contents of <C:/_qclwr/_inputfiles/corp_ice-gb.txt>.

```
[5]·"a·hundred"
[6]·"problem"
```

The tabulation, sorting, and saving of the results is done as before. Hopefully, however, while the above is a workable solution, you immediately noticed a shorter way of doing this once current.corpus.file.2 has been generated...

THINK

BREAK

```
>·words<-gsub(".*{([^}]+)}",·"\\1",·current.corpus.file.2,·perl=T)¶
```

This replaces the whole line by only the back-referenced item between the curly brackets. Another solution worth exploring uses Gabor Grothendieck's package gsubfn. You can use immediately after the corpus file has been loaded:

```
>·library(gsubfn)¶
>·words<-unlist(strapply(current.corpus.file,·"{([^}]*)}",·c,·perl=T))¶
```

If you want to process several corpus files, you could do all this in a simple for-loop by using a vector whole.corpus to which the content of current.corpus.file.4 is successively appended and, once the loop is exited, use table to get the frequency list. Alternatively, you might want to store all the frequency lists for the files in a list and then merge them with the solution to Question 10 in Exercise box 3.1.

4.1.7 A Frequency List of Word Pairs of an Annotated Corpus (with One Word Per Line)

If, instead of a frequency list of *words*, you were interested in a frequency list of *word pairs*, you might not like the interim result of Section 4.1.6 because if you now simply generated word pairs by collapsing neighboring words in the vector, you would again collapse all last words of a sentence with the first words of the following sentences, a strategy we deliberately avoided in Section 4.1.5. Thus, you can rearrange the corpus file in such a way that the data are in lines that correspond to what the corpus annotators considered to be parse units.

We immediately load the package gsubfn and the corpus file as well as merge all corpus lines into one string, but choose a separator that does not already occur in the file and, thus, allows you to still identify line breaks:

```
>•library(gsubfn)¶
>•current.corpus.file<-scan(choose.files(),•what="char",•sep="\n")•#•load•
  the•corpus•file¶
>•current.corpus.file.2<-paste(current.corpus.file,•collapse="___")•#•merge•
  all•corpus•lines•into•a•single•string¶
```

Then, you split up this one string at the boundaries of parse units, which will be the boundaries across which you do not want to combine words into pairs. To that end, you split up the long single string again, but the separator is now "___PU" because "PU" at the beginnings of lines indicates the beginning of sentences or parse units and you just put "___" in front of every line and, thus, also in front of every "PU":

```
>•current.corpus.file.3<-unlist(strsplit(current.corpus.file.2,•"___PU,"))•
  #•split•up•this•string•again•at•the•boundaries•of•parse•units¶
```

Do you now wonder why we put in the three underscores? And why we did not just paste the file together and split it up at every "PU"? If yes, think about it for a moment . . .

THINK

BREAK

We did this because there may be instances of "PU" in the corpus as *words* rather than as *annotation*. For example, PU is the symbol for plutonium. If you pasted all lines together without the three underscores preserving the information where line breaks were and then split up the whole string at every PU, you may have split up the data at positions where no line break showed that PU was annotation meaning "parse unit". This strategy of course requires that you also know that there are not already instances of "___PU" in the data, which is obviously the case here.

As the next step, you disregard all elements that do not contain words, i.e. that do not contain "{" or "}":

```
>•current.corpus.file.4<-grep("[{}]",current.corpus.file.3,•perl=T,•value=T)•
  #•retain•only•those•elements•containing•words¶
```

and then use `strapply` in the same way as in the previous section to get at the words within the curly brackets:

```
>•current.corpus.file.5<-strapply(current.corpus.file.4,•"{([^}]*)}",•c,•
  perl=T,•backref=-1)¶
```

This way, we restrict our generation of word pairs to each parse unit and avoid the formation of word pairs from different sentences; we thus go on as in Section 4.1.5:

```
>•output.file<-choose.files()¶
#•and•so•on•.•.•.•try•it•out•before•you•check•the•answer•key¶
```

You should now do "Exercise box 4.1: Frequency lists" . . .

These are certainly not all the ways in which you can generate or use frequency lists, but this section covered several of the most frequent applications, and the case study assignments from the companion website will take up many of the above issues to deepen your understanding of the contents of this section. The next section will be concerned with concordancing.

For further study/exploration: Think about how you would process several files to generate and save a separate frequency list of each file, a useful feature that in WordSmith Tools 4 is called batch frequency lists.

4.2 Concordances

4.2.1 A Concordance of an Unannotated Text File

Again, we begin with the simplest case of a single unannotated text file, the file <C:/_qclwr/_input files/corp_gpl_short.txt>. We first read in the file as before:

```
>•textfile<-scan(choose.files(),•what="char",•sep="\n")¶
Read•9•items
```

The simplest way to generate a concordance involves grep as in this example:

```
>•grep("the",•textfile,•ignore.case=T,•value=T,•perl=T)¶
[1]•"The•licenses•for•most•software•are•designed•to•take•away•your"
[2]•"freedom•to•share•and•change•it.•By•contrast,•the•GNU•General•Public"
[3]•"software--to•make•sure•the•software•is•free•for•all•its•users.•This"
[4]•"General•Public•License•applies•to•most•of•the•Free•Software"
[5]•"Foundation's•software•and•to•any•other•program•whose•authors•commit•to"
[6]•"using•it.•(Some•other•Free•Software•Foundation•software•is•covered•by"
[7]•"the•GNU•Library•General•Public•License•instead.)•You•can•apply•it•to"
```

You can store the result of such a search as a vector and then into a file (for example, <C:/_qclwr/_outputfiles/04–2–1_output-a.txt>) using what we already know:

```
>•conc.the.1<-grep("the",•textfile,•ignore.case=T,•value=T)¶
>•cat(conc.the.1,•file=choose.files(),•sep="\n")¶
```

Two things are worth noting here. The first is that R of course again does not know that *the* is a word. Thus, while a human user may consider it obvious that he only wants all occurrences of *the* as a word, for R this is not obvious. Accordingly, the result contains all occurrences of the letter sequence *the* irrespective whether a particular match is really an instance of the word *the* or whether *the* is part of a larger sequence (as in match 5 above). How could we handle this problem? At first sight, you may think of using the notion of non-word characters. However, this runs into problems:

```
>•grep("\\Wthe\\W",•textfile,•ignore.case=T,•value=T)¶
[1]•"freedom•to•share•and•change•it.•By•contrast,•the•GNU•General•Public"
[2]•"software--to•make•sure•the•software•is•free•for•all•its•users.•This"
[3]•"General•Public•License•applies•to•most•of•the•Free•Software"
```

As you can see, this fails to match the elements where the match is at the beginning of the character string because then "the" is not preceded by a non-word character. However, we know a metacharacter that works better: word boundaries.

```
>•conc.the.1<-grep("\\bthe\\b",•textfile,•ignore.case=T,•value=T)¶
>•conc.the.1¶
[1]•"The•licenses•for•most•software•are•designed•to•take•away•your"
[2]•"freedom•to•share•and•change•it.•By•contrast,•the•GNU•General•Public"
```

```
[3]•"software--to•make•sure•the•software•is•free•for•all•its•users.•This"
[4]•"General•Public•License•applies•to•most•of•the•Free•Software"
[5]•"the•GNU•Library•General•Public•License•instead.)•You•can•apply•it•to"
```

We now have the "right" output (compare <C:/_qclwr/_outputfiles/04-2-1_output-b.txt>). The second point is that, unlike most concordance programs, R does not output a separate line for each match. Thus, if there is more than one match in a line, you will only get the line retrieved just once. On the one hand, this is not so good because usually additional coding in spreadsheets, etc. is easier when each case has its own line. On the other hand, this is good because it guarantees that you will always get to see all examples in their proper context. However, the output can be customized so as to make it easier to continue working with it. One way of achieving this is to make sure that each line in which a match is found is split up into at least three parts, viz. one with the context preceding the match, one with the match, and one with the context following the match. The easiest way to achieve this is using a function to replace every match by the sequence of a tab stop, the (back-referenced) match, and another tab stop. The function used to achieve this before saving the result into <C:/_qclwr/_outputfiles/04-2-1_output-c.txt> is gsub.

```
>•conc.the.2<-gsub("\\b(the)\\b",•"\t\\1\t",•conc.the.1,•ignore.case=T)¶
>•cat("PRECEDING_CONTEXT\tMATCH\tSUBSEQUENT_CONTEXT",•conc.the.2,•
   file=choose.files(),•sep="\n")•#•output¶
```

Stop reading and think for a moment about why we use a back-reference rather than just "\tthe\t" as a replacement pattern.

THINK

BREAK

We use a back-reference because using "\tthe\t" as a replacement would automatically change all matches to lower case, so if we want to preserve the *exact* match, we must instruct R to store it and use it again.

If you now open this output file with a spreadsheet software in the way discussed above in Section 4.1.1, you will see that the file is organized as desired and can be further fine-tuned and processed as in Figure 4.4.

While for many purposes, this output may actually already be sufficient, on other occasions you might want to generate a different kind of output. For example, a downside of the above output is that, for some matches, there may be no preceding or subsequent context within the line. There are different quick fixes to address this issue. The simplest consists of just adding the lines before and after the match to the output. Let us look at this in more detail. First, we retrieve the lines in which matches occur:

```
>•(matches.the.1<-grep("\\bthe\\b",•textfile,•ignore.case=T,•value=F))¶
[1]•1•2•4•5•8
```

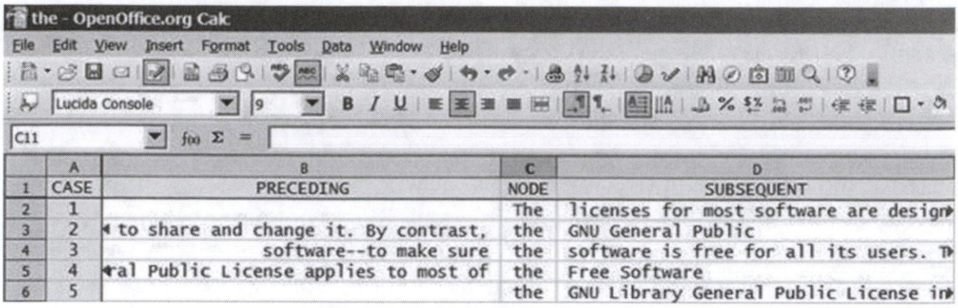

Figure 4.4 A concordance file imported into OpenOffice.org Calc, format 1.

This should also alert you to the fact that not all matches have a previous or subsequent line—cf. the first in our present example—so we have to make sure that the output is adjusted accordingly. More explicitly, we need the following output (the lines with the match are underlined):

- lines 1 and 2 for the first match;
- lines 1, 2, and 3 for the second match;
- lines 3, 4, and 5 for the third match;
- lines 4, 5, and 6 for the fourth match;
- lines 7, 8, and 9 for the fifth match.

One straightforward dirty way of doing this is a for-loop with as many iterations as we have matches. We specify an output file (<C:/_qclwr/_outputfiles/04-2-1_output-d.txt>), print a column header, and then begin with a for-loop:

```
>•output.file<-choose.files()¶
>•cat("PRECEDING_CONTEXT\tMATCH\tSUBSEQUENT_CONTEXT\n",•file=output.file)¶
>•for•(i•in•matches.the.1)•{•#•access•each•match¶
```

For each match, we access (i) the line immediately preceding the position of the current match, (ii) the current match itself, and (iii) the immediately subsequent line:

```
+••••previous.context<-textfile[(i-1)]•#•access•the•immediately•preceding•
    vector•element;•if•there•is•none,•this•evaluates•to•an•empty•character
    vector¶
+••••current.match<-textfile[i]•#•access•the•current•match¶
+••••subsequent.context<-textfile[(i+1)]•#•access•the•immediately•following•
    vector•element;•if•there•is•none,•this•evaluates•to•NA•(not•available)¶
```

We then output the contexts and the match separated by tab stops (to ease further processing in a spreadsheet software). Note that we print into the output file defined above and that we set append=T so that the output from later matches is added to that of earlier matches and does not overwrite it:[6]

```
+••••cat(previous.context,•"\t",•current.match,•"\t",•subsequent.context,•
  "\n",•file=output.file,•append=T)•#•output¶
+•}•#•end•of•for:•access•each•match¶
```

A similar quick fix is possible if your search expression is just one word or a part of a word and would require to first change the way the corpus is represented in R. At the moment, each line is one element of the character vector `textfile` but you could of course also have each word stored as an element of a character vector. A first though not particularly elegant way of getting this corpus is the following:

```
>•textfile.2<-unlist(strsplit(textfile,•"\\W+"))¶
```

This may often be sufficient but you must be aware that `textfile.2` no longer has punctuation marks anymore because these were used by `strsplit`. If you do not want to lose the punctuation marks, however, you can first surround every non-word character with spaces and then use all spaces—the ones that surrounded words in the first place and the new ones now surrounding everything else—to split up the character strings:

```
>•textfile.2<-gsub("(\\W)",•"•"•\\1•",•textfile,•perl=T)¶
>•textfile.3<-unlist(strsplit(textfile.2,•"•+"))¶
>•tail(textfile.3)¶
[1]•"to"••••••"your"•••••"programs"•",".•••••••••"too"••••••"."
```

Now, you can search for your search expression:

```
>•(matches.the<-grep("\\bthe\\b",•textfile.3,•ignore.case=T))¶
[1]••1•22•44•61•90
```

Now we have the positions of all matches. But how do we turn that into a nice output? Well, we might try to use a for-loop to output every match together with a fixed number of elements before and after the match. If, for example, we would like to obtain the second match with its preceding and subsequent six words, then this line would give you the desired output:

```
>•textfile.3[(matches.the[2]-6):(matches.the[2]+6)]¶
```

What does this line do? It accesses parts of `textfile.3`. Which parts? First, R accesses the second position of the vector with the positions of the matches (`matches.the[2]`), which amounts to 22. Second, R computes the range of parts to be outputted, namely 16:28 (22 − 6:22 + 6). This range of values is then outputted. For expository reasons, this could be written very redundantly as follows:

```
>•current.match<-2¶
>•desired.span<-6¶
>•current.position.in.corpus<-matches.the[current.match]•#•22¶
>•beginning.of.range<-current.position.in.corpus-desired.span•#•16¶
>•end.of.range<-current.position.in.corpus+desired.span•#•28¶
>•range<-beginning.of.range:end.of.range;•range¶
[1]•16•17•18•19•20•21•22•23•24•25•26•27•28
>•textfile.3[range]•#•output¶
•[1]•"change"•••"it"•••••••"."•••••••••"By"•••••••"contrast"•","
•[7]•"the"•••••••"GNU"•••••••"General"••"Public"•••"License"••"is"
[13]•"intended"
```

Unfortunately, this is not the end of the story: this approach doesn't work for the first match: the first subtraction, matches.the[1]−6, results in a negative subsetting value: −5:

```
>•textfile.3[(matches.the[1]-6):(matches.the[2]+6)]¶
Error:•only•0's•may•be•mixed•with•negative•subscripts
```

We therefore need to adjust our output function so that the minimum subsetting value cannot be smaller than 1. However, we do not really need that much about the maximum subsetting value: if you pick a value larger than the length of the text vector, 108 in this case, then R returns NA, meaning that part of the text vector does not exist. In R, you could do this:

```
>•beginning.of.range<-max(1,•(matches.the[1]-6))•#•1:•the•max•rules•out•
   indices•≤•0¶
```

Now we are just about done. The only issue that remains is that we now must access each match and compute the beginnings and ends of its ranges automatically (and not, like above, by entering the index of the match manually). We can use a for-loop to go through all the results automatically and print them all into one file.

Let me show how this is done in very few lines and, since this is a rather complex snippet, I will discuss it in detail. First, you must choose an output file:

```
>•output.file<-choose.files()•#•choose•an•output•file¶
```

Then, you start a for-loop that will be executed as many times as there are matches in the concordance.

```
>•for•(i•in•matches.the)•{•#•go•through•all•matches•in•the•file¶
+••••cat(file=output.file,•append=T,¶
```

Then, you begin the output process with the instruction to write the output into the user-defined output file and continually append output to that file (rather than overwriting it) each time the loop is processed.

```
+•••••••••textfile.3[max(0,•i–6):max(0,•i–1)],¶
```

Now you begin to access a part of the corpus. Note that we use `max(0,•i–6)`, not `max(1,•i–6)` as above. This is because in the first loop the script accesses the position of the first match in the corpus, which is 1. Still in this line, the script then tries to look back from the first position. However, since the first match occurs at position 1, `max(0,•i–6)` returns a 0 (because 0 > –5). Similarly after the colon, `max(0,•i–1)` returns a 0. Thus, R accesses and prints `textfile.3[0:0]`, which amounts to " ". This is because the first position of `textfile.3` that actually contains an element is 1, so we must not output anything for the preceding context of the first word:

```
+•••••••••"\t",¶
```

Then, you print a tab stop to the file to separate the preceding context from the match, the match itself, and another tab stop.

```
+•••••••••textfile.3[i],¶
+•••••••••"\t",¶
```

So far, the output file looks like Figure 4.5, with one tab stop both before and after "The".[7]

The	¶

Figure 4.5 Temporary content of the concordancing output 1.

```
+•••••••••textfile.3[(i+1):min(i+6,•length(textfile.3)+1)],¶
```

This line looks at the words after the match. The range begins with the position after the match, which is either the word following the match or NA, if the match is the last word of the file and there is no such word. This range ends with the minimum of either six words after the match or, when the match is close to the end, the length of the text file plus one. Thus, R effectively determines whether it can output as many elements as the user-defined span (here: 6) would require or whether the end of the corpus file is reached first: in the case of the first iteration, `min(length(textfile.3)+1,•i+6)` output 7.

```
+••••••••"\n")¶
+•}¶
```

Then, the script ends the line in the output file with a newline; this is the result of executing the loop for the first time; cf. Figure 4.6.

```
The    licenses for most software are designed ¶
```

Figure 4.6 Temporary content of the concordancing output 2.

After that, the script proceeds through the loop as many times as there are matches in the corpus to produce the complete concordance.

Now that this process has been completed, a comment is in order concerning the actual contents of the file. Above, we said that the file contains words and tab stops. However, if you open the output file, you will see that it actually contains more, namely (seemingly irrelevant) spaces before and after tab stops and before the newline. The first line, for example, looks like Figure 4.7, where the small dots and the arrows again represent spaces and tab stops respectively.

```
→    ·The·  →  ·licenses·for·most·software·are·designed·¶
```

Figure 4.7 Temporary content of the concordancing output 3 (higher resolution).

Where do these come from? They are a result of our not having specified a separator character for cat so R used the default setting for this argument, a space (cf. Section 3.2.4). Thus, all the elements that are processed within one call of cat are separated with tab stops. Could we have avoided that? Yes, we could have (by specifying a separator character), but the R default setting is actually quite useful here because it allows us to print all the words from the preceding context in one go since we can trust cat to put spaces between them. Thus, if we had written cat(. . .,•sep=""), we would have avoided the spaces, but the context words would not have been separated properly, as is shown in Figure 4.8.

```
→    The    →    licensesformostsoftwarearedesigned¶
```

Figure 4.8 Hypothetical content of the concordancing output 2.

Given the results with the spaces, I recommend cleaning up the file before you use it any further. For example, you may want to reduce sequences of spaces and tab stops (in either order) to just a tab stop, and/or you may want to delete leading and trailing spaces at the beginnings and ends of lines. Once this kind of cleaning up has been completed, the file can be loaded into a spreadsheet software for additional annotation and processing; cf. Figure 4.9.

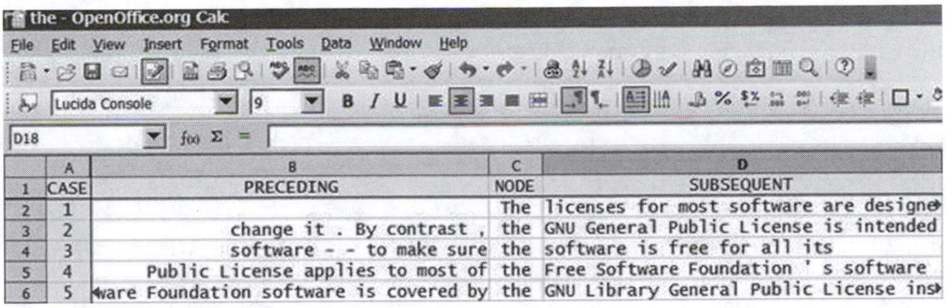

the - OpenOffice.org Calc

	A	B	C	D
1	CASE	PRECEDING	NODE	SUBSEQUENT
2	1		The	licenses for most software are designe▸
3	2	change it . By contrast ,	the	GNU General Public License is intended
4	3	software - - to make sure	the	software is free for all its
5	4	Public License applies to most of	the	Free Software Foundation ' s software
6	5	◂ware Foundation software is covered by	the	GNU Library General Public License ins▸

Figure 4.9 A concordance file imported into OpenOffice.org Calc, format 2.

4.2.2 A Simple Concordance from Files of a POS-tagged (SGML) Corpus

On many occasions, we will want to make use of the fact that our corpus is already annotated in some way(s). Given the wide distribution of the SGML-annotated version of the BNC, we will therefore again look at an example involving this corpus. Let us assume we would like to retrieve all cases of the word *right* when tagged as an adjective in the files of the BNC from the register of spoken conversation. Let us also assume we do not just want the matches as such, but also the exact locations of the matches in the output. As you will notice, we will be able to recycle many lines of code we used in Section 4.1.4 above. First, we choose the files we want to search and generate a vector for the results:

```
>•rm(list=ls(all=T))¶
>•corpus.files<-choose.files()•#•choose•all•corp_bnc_sgml•files¶
>•all.matches.right<-vector()¶
```

Then, we use a `for`-loop to access every file in our vector `corpus.files` and output the name of the corpus files currently processed to monitor progress, etc. Each of the files is imported into a vector called `current.corpus.file`, using `scan` in the by now familiar way:

```
>•for•(i•in•corpus.files)•{¶
+••••cat(i,•"\n")•#•output•a•'progress•report'¶
+••••current.corpus.file<-scan(i,•what="char",•sep="\n",•quiet=T)¶
```

Next, we must check whether the file is one of those containing spoken conversation. As you can look up in the documentation of the BNC, each file has a header line (starting with "<teiHeader") providing information about the source of the file, the speakers recorded, the register of the file, some key words, etc. The register information can be found in the header line between a beginning and ending tag labeled "classCode".

For further study/exploration: `scan` provides a more efficient way of performing this register check because `scan` allows you to specify the number of lines of the file that are read. Since, in the BNC, the header line is always the second line, you could just read in the first two lines and only access more of the file when the file is of the desired kind

Thus, we check whether the header line contains the register code for spoken conversation and only continue with searching this file if that is the case:

```
+••••spoken.conversation<-grep("<teiHeader•type.*S•conv</classCode>",•
  current.corpus.file,•perl=T)¶
+••••if•(length(spoken.conversation)==0)•{•next•}¶
```

When grep has been performed with value=F—the default—you can also test whether there are any matches with the following: if•(any(spoken.conversation))•{•next•}¶, and any non-zero number will make the expression true. If the file is of the right kind, we again reduce the current corpus file to only those lines that have a BNC-style sentence number to avoid the header, utterance tags, etc.

```
+••••current.sentences<-grep("<s•n=",•current.corpus.file,•perl=T,•value=T)¶
```

However, this time we will not delete the sentence numbers because we want to retain them so that the output can tell us where the match(es) in each file were found.

Let us now craft the regular expression for the matches. Let's assume you do not only want matches where *right* is tagged as an adjective proper (the tag for an adjective in the base form is "AJ0"), but also when *right* is tagged with a so-called portmanteau tag, a combination of two tags, which the tagger assigns if "it is not sure which tag is the right one". There is one portmanteau tag relevant to our present example, "<w•AJ0-AV0>", so if you want to include these cases in your results—because you want to decide unclear cases for yourself—this is what you could do:

```
+••••current.matches.right<-grep("<w•AJ0(-AV0)?>right\\b",•current.sentences,•
  ignore.case=T,•perl=T,•value=T)¶
```

However, there is another small twist. You would typically also want the matches' exact locations in the output. While the sentence numbers are part of the retrieved lines, the names of the files in which the matches occur are not. Thus, you first check whether the search in the current file has returned any matches,

```
+••••if•(length(current.matches.right)==0)•{•next•}¶
```

. . . and if it has, you attach the short versions of the file names with paste and separate them from the rest of the output by a tab stop,

```
+••••current.matches.right.2<-paste(basename(i),•current.matches.right,•
  sep="\t")¶
```

. . . and then store the string containing the file name and the match for later output, and end the loop.

```
+••••all.matches.right<-append(all.matches.right,•current.matches.right.2)¶
+•}•#•end•of•for:•go•through•all•files¶
```

Now the only thing that remains is preparing the data for the output and actually outputting them. As before, we could just replace the matches by matches with surrounding tab stops and print everything into file such as <C:/_qclwr/_outputfiles/04-2-2_output-a.txt>:

```
>•all.matches.right.2<-gsub("(<w•AJ0(-AV0)?>right)\\b",•"\t\\1\t",•
   all.matches.right,•perl=T)¶
>•cat(all.matches.right.2,•file=choose.files(),•sep="\n")¶
```

Note how the substitution is done. The first set of parentheses delimits the search expression, i.e. the tag and the word "right". The second set of parentheses within the first serves to define the optional portmanteau tag. The replacement then involves inserting a tab before and after the tag-word pair (which is back-referenced as "\\1").

Further ways of customizing the output are conceivable and you should explore them all. For example, you might wish to put the sentence number into its own column, in which case you could retrieve the sentence number tag and replace it by just the back-referenced sentence number followed by a tab stop:

```
>•all.matches.right.3a<-gsub("<s•n=\"(\\d+)\">",•"\\1\t",•
   all.matches.right.2,•perl=T)¶
```

Also, you may want to decide to delete all POS and punctuation mark tags, but retain all others. All tags are surrounded by "<" and ">2", but the tags to be removed (i) all start with "<w" or "<c" followed by a space and (ii) have either three characters or three characters, a minus, and three additional characters. Thus:

```
>•all.matches.right.3b<-gsub("<[wc]•(...|...-...)>",•"",•
   all.matches.right.3a,•perl=T)¶
```

Alternatively, you may want to delete all tags other than POS, punctuation mark tags, and sentence numbers. An elegant and extremely useful way would involve negative lookahead: this line looks for everything between angular brackets that does *not* match word and punctuation tags as well as the sentence numbers. Again, you can now clean up this vector and output it.

```
>•all.matches.right.3c<-gsub("<(?![wcs]•(...|.......|n=.*?)).*?>",•"",•
   all.matches.right.3a,•perl=T)¶
```

Lastly, you may want be to delete leading and trailing spaces, i.e. spaces before and after tab stops, to facilitate sorting procedures in OpenOffice.org Calc. You can do this as before:

```
>•all.matches.right.3d<-gsub("•*\t•*",•"\t",•all.matches.right.3b,•perl=T)¶
```

If you look at the result (especially the OpenOffice.org Calc version in <C:/_qclwr/_outputfiles/04 –2–2_output-a.ods>), you will notice something that was mentioned above before and that is not so nice: there are lines/rows with up to three matches. For instance, row 35 contains two matches, which makes a subsequent annotation of each match cumbersome. In this section, I will therefore do two important things. First, I will introduce a way how this issue can be addressed, which is a little complex, but extremely powerful—once you have mastered this, you can customize output extremely flexibly. Second and to your relief, I will provide you with a function that implements that way and makes it much much easier for you to get nicely formatted results.

This approach involves the function gregexpr. To make the following easier, let us for now use only the first of the above corpus files, <corp_bnc_sgml_1.txt>:

```
>•rm(list=ls(all=T))¶
>•corpus.file<-scan(choose.files(),•what="char",•sep="\n",•quiet=T)¶
>•sentences<-grep("<s•n=",•corpus.file,•perl=T,•value=T)¶
```

Now, how can gregexpr help us? Let us first perform the search and look at its output. Before we simply apply gregexpr to sentences, recall that the output of gregexpr was a list with as many elements as the searched vector had elements, because if there was no match, gregexpr would still return −1. Thus, it is usually more efficient to use grep to narrow down the corpus (files) to only those lines that do contain matches first, and only then use gregexpr on the result of grep:

```
>•lines.with.matches<-grep("<w•AJ0(-AV0)?>right\\b",•sentences,•
  ignore.case=T,•perl=T,•value=T)¶
>•matches<-gregexpr("<w•AJ0(-AV0)?>right\\b",•lines.with.matches,•
  ignore.case=T,•perl=T)¶
>•matches[34:36]¶
[[1]]
[1]•199
attr(,"match.length")
[1]•12

[[2]]
[1]•223•283
attr(,"match.length")
[1]•16•12
```

```
[[3]]
[1] 74
attr(,"match.length")
[1] 12
```

You can see how `gregexpr` finds both matches in `matches[35]`. But how can we use this fairly unwieldy input to get the exact matches? Since `gregexpr` returns positions within strings, it's only natural to use `substr`. As you probably recall, `substr` requires three arguments: the vector from whose elements you want substrings, the starting positions of the substrings, and the end positions of the substrings. Let us construct how `substr` must be used here.

First, the vector from which we want substrings. We want substrings from the vector `lines .with.matches`, but note what exactly it is we want. For example, we want one substring from `lines.with.matches[34]`, but we want two substrings from `lines.with.matches[35]`. The first argument to `substring`, thus, must make sure that you can access every element of `lines.with .matches` as many times as it contains matches. How do we find how many matches each element of `lines.with.matches` has? Simple, we already did that above:

```
> sapply(matches, length)¶
  [1] 1 1 1 1 1 1 1 1 1 1 1 1 1 1 1 1 1 1 1 1 1 1 1 1 1 1 1 1
 [29] 1 1 1 1 1 1 2 1 1 1 1 1 1 1 1 1 1 1 1 1 1 1 1 1 1 1 1 1
 [57] 1 1 1 1 1 1 1 1 1 1 1 3 1 1 1 1 1 2 1 2 1 1 1 1 1 1 1 2
 [85] 1 1 1 1 1 1 2 1 1 1 1 1 1 1 1 1 1 1 1 1 1 1
```

Now, in order to have each element of `lines.with.matches` as many times as it contains matches, we use the function `rep`, whose first argument is what is to be repeated, and whose second argument is how often something is to be repeated. The crucial point is that `rep` can be applied to vectors with more than one element. The following line returns the 4 once, the 5 twice, and the 6 three times:

```
> rep(4:6, 1:3)¶
[1] 4 5 5 6 6 6
```

Thus, the first argument to `substr` will be:

```
> source<-rep(lines.with.matches, sapply(matches, length))¶
```

The second argument to `substr` is easy—it's the starting points of the matches:

```
> starts<-unlist(matches)¶
```

The third argument to substr requires a bit of addition. Remember: gregexpr returns the *starting positions* and the *lengths* of matches while substr needs *starting positions* and *end positions*. Thus, we need to do a little addition to infer the end position from the starting position and the length. We take the starting position, add the length of the match, and subtract 1:

```
>•stops<-starts+unlist(sapply(matches,•attributes))-1¶
```

If you now put it all together, you can write the following to inspect all exact matches:

```
>•precise.matches<-substr(source,•starts,•stops)¶
```

This is already very useful because this is how we basically always expected searches to work—we only get what we looked for. But we also want the context—can you see how we get that?

THINK

BREAK

Although not easy to come up with, but it is easy to understand . . .

```
>•delimited.matches<-paste(¶
•••substr(source,•1,•starts-1),•"\t",•#•from•beginning•of•line•to•match¶
•••substr(source,•starts,•stops),•"\t",•#•the•match¶
•••substr(source,•stops+1,•nchar(source)),•#•from•match•to•end•of•line¶
•••sep="")¶
>•cat("PRECEDING_CONTEXT\tMATCH\tSUBSEQUENT_CONTEXT\n",•delimited.matches,•
  sep="\n",•file=choose.files())•#•<04-2-2_output-b.txt>¶
```

You just request the parts of each line with its match separately. As you can now see, there were 105 lines with matches, but delimited.matches now contains 112 lines, and each line contains only one match; look especially at the rows with cases 35 and 36 in <C:/_qclwr/_outputfiles/04–2–2_output-b.ods>. Of course, you could put the file names, etc. in as well.

For further study/exploration:

- on how to get the beginning and starting points of strings different from how we did it: ?substring.location¶ in the package Hmisc
- on how to perform replacement operations on vectors: ?sedit¶ in the package Hmisc
- on how to test whether strings are numeric and whether they only contain digits respectively: ?numeric.string¶ and ?all.digits¶ in the package Hmisc

I am pretty sure you are now exhausted enough and ready for the good news . . . The good news is that in your folder <C:/_qclwr/_scripts/> there is a script whose contents you can copy and paste into R (or load it into R with the function source i.e. source("C:/_qclwr/_scripts/exact_ matches.r")¶—or, on Windows, drag and drop it onto the R console), and this function does all this—and more—for you. The function takes four arguments:

- a search pattern;
- the vector to be searched;
- pcre=T, which is the same as perl=T (the default), or pcre=F, if you don't want to use perl-compatible regular expressions;
- case.sens=T, which means the same as ignore.case=F (the default), or case.sens=F, if you want to perform case-insensitive searches.

Thus, instead of all the things above, you can now just write this:

```
>•out<-exact.matches("<w•AJ0(-AV0)?>right\\b",•sentences,•pcre=T,•
   case.sens=F)¶
>•sapply(out,•head)¶
```

This function basically uses the above approach to create a list with four components, which you can look at by entering out[[1]], etc.

- component [[1]]: a vector called "exact matches", which gives you the exact matches (without context);
- component [[2]]: a vector called "locations of matches", which gives you the positions where matches were found in the input vector; for example, one match was found in line 235, two matches were found in line 562, etc.;
- component [[3]]: a vector called "lines with delimited matches": the complete lines with tab-delimited matches (i.e. what we outputted into <04-2-2_output-b.txt>);
- component [[4]]: a vector that repeats the parameters with which exact.matches was called.

While this function will certainly make your life very much easier and several sections below will use it, I strongly recommend that you nevertheless try to understand the logic underlying its code so that you later are able to customize your output according to your own wishes.

4.2.3 More Complex Concordances from Files of a POS-tagged (SGML) Corpus

In terms of regular expression used for searching, though certainly not otherwise, the preceding example was simple and did not really go beyond what can be done with unannotated text files. The main improvement over the search expression from Section 4.2.1 was to include one simple tag. In this section, we will look at a few examples involving the BNC where the power of regular expressions in R will become much more obvious. In the interest of brevity, we will restrict ourselves to the retrieval of the matches from one file (<C:/_qclwr/_inputfiles/corp_bnc _sgml_1.txt>) and will do little to customize the output even if one would usually do this in practice. Let's prepare the data:

```
>•rm(list=ls(all=T))¶
>•source("C:/_qclwr/_scripts/exact_matches.r")¶
>•current.corpus.file<-scan(choose.files(),•what="char",•sep="\n",•quiet=T)•
  #•access•the•corpus•file•mentioned•above¶
>•current.sentences<-grep("<s•n=",•current.corpus.file,•perl=T,•value=T)•
  #•choose•only•the•lines•with•sentence•numbers¶
```

Then, we get rid of the things that complicate our search, viz. annotation that does not identify POS tags (for words and punctuation marks) and sentence numbers. Thus, we just recycle our negative lookahead expression from above:

```
>•current.sentences<-gsub("<(?![wcs]•(...|....-...|n=.*?)).*?>",•"",•
  current.sentences,•perl=T)¶
```

4.2.3.1 Potential Prepositional Verbs with in

Our first small case study is concerned with identifying potential instances of prepositional verbs with *in* (I am using the terminology of Quirk et al. 1985). While transitive phrasal verbs allow the separation of their verb and their (adverbial) particle, prepositional verbs in English usually require that the verb immediately precedes its preposition. Thus, our regular expression becomes only slightly more complex. Let us assume we want to retrieve instances of potential prepositional verbs with the preposition *in* from a corpus file. Our regular expression must consider three parts: (i) the expression matching the tag for the verb of the prepositional verb, (ii) the regular expression matching the actual verb, and (iii) the expression matching the tag of the preposition and the preposition *in* itself. As to the first, we want to find verbs, but not any kind of verb. For example, we usually would not expect a prepositional verb to have *to be* or a modal verb as its verbal component. Also, *to do in* and *to have in* are rather unlikely as prepositional verbs at least. Since forms of the verbs *to be, to do,* and *to have* have their own tags in the BNC ("<•VB.>", "<w•VD.>", and "<w•VH.>" respectively), we can use the BNC tags for lexical verbs, which begin with "VV". However, we should maybe not just match "<w•VV.>", because there may be portmanteau tags in which the "VV." part is followed by some other second part. So the first part of our regular expression is "<w•(VV.|VV.-...)>".

As to the last part, this one is easy. We dealt with this in Section 4.2.2 and can thus simply write "<w•PRP>in\\b". But the part that is more interesting is the intermediate one, which must match the verb following the tag. One way to go is this. Since we know that the verb can only extend to the beginning of the next tag—"<", that is—we can simply define the verb as the shortest possible sequence of characters that are not "<" but only until the next "<". Thus, the next part of the regular expression is "[^<]*?". Thus, the whole line of code is this:

```
>•current.matches<-exact.matches("<w•(VV.|VV.-...)>[^<]*?<w•PRP>in\\b",•
  current.sentences,•pcre=T,•case.sens=F)¶
```

If the third component of this list gets printed into a file, you get the results in <C:/_qclwr/_output files/04–2–3–1_output.txt>.

4.2.3.2 Preposition Stranding in Questions

Our second small case study involves preposition stranding in English *wh*-questions, i.e. constructions in which the formation of a *wh*-question leaves a preposition stranded at the end of the question and which is often frowned upon from a prescriptive point of view. Examples for preposition stranding are given in (7) and (8), with the stranded and the pied-piped versions in a. and b. respectively.

(7)   a. Who$_i$ did John give the book [$_{PP}$ to t$_i$]?
        b. [$_{PP}$ To whom]$_i$ did John give the book t$_i$?

(8)   a. Who$_i$ did you see a picture [$_{PP}$ of t$_i$]?
        b. [$_{PP}$ Of whom]$_i$ did you see a picture t$_i$?

We match something that is tagged as a *wh*-pronoun somewhere in the line and that is later followed by a preposition which in turn is immediately followed by a question mark:

```
> current.matches<-exact.matches("<w (DTQ|PNQ)>.*?<w PRP>[^<]*?<c PUN>\\?", 
  current.sentences, pcre=T, case.sens=F)¶
```

If the third component of this list gets printed into a file, you get the results in <C:/_qclwr/_out putfiles/04–2–3_2_output.txt>. Of course, this search will require much manual editing since the distance between the *wh*-pronoun and the question mark can be huge, allowing for many different kinds of false hits. It would be nice if the length of the intervening sequence could be restricted.

For further study/exploration: Gries (2002b) and Hoffmann (2006) and the references cited there

You should now do "Exercise box 4.2: Other advanced regular expressions" ...

4.2.3.3 Potential Verb–particle Constructions

This example involves one of the most complex regular expressions in this book: we will retrieve constructions with transitive phrasal verbs from the BNC. As mentioned in Section 4.2.3.1, transitive phrasal verbs are characterized by the fact that they allow two different constituent orders: one in which a verb and an adverbial particle are directly followed by the direct object of the transitive phrasal verb (cf. the a. examples in (9) and (10)), and one in which the direct object intervenes between the verb and the particle (cf. the b. examples in (9) and (10)).

(9)   a. John gave away all his old books.
        b. John gave all his old books away.

(10) a. Mary gave away the secret.
        b. Mary gave the secret away.

These examples indicate what is the trickiest aspect of this regular expression: We must write a regular expression that can accommodate different numbers of words between the verb and the particle. In addition, our regular expression must also allow each of these words to come with a normal or a portmanteau tag. Finally, we may not want to allow any of the intervening words to be verbs themselves.

For now, let us restrict the number of words between the verb and the particle to the range from 0 (i.e. the a examples above) to 5. We again begin only after current.sentences has been cleaned by considering the different parts our regular expression will need: We need:

- the tag of the verb;
- the verb;
- the material that may intervene between the verb and the particle;
- the particle (tagged as "<w·AVP>" in the BNC).

The first two parts are not so difficult given the previous sections: Restricting our attention to lexical verbs only, the tag of the verb is simply defined as "<w·(VV.|VV.-. . .)>". The verb after this tag is described as "[^<]*?". The fourth part is also easy: we want the tag of the particle plus the particle: "<w·AVP>[-a-z]*". Note precisely what this expression does: it looks for occurrences of the tag, followed by the longest number of characters ("*") that are hyphens, letters, or a space; note in particular that if we want to include a minus/hyphen in character class, we must put it first to make sure it is not interpreted as a range operator (as in "[a-z]"). This way we include the possibility that we have a hyphenated particle (such as *in-out*, *mid-off*, or *mid-on*) or a particle with a space in it in case *back up* in *He brought it back up* was tagged as one particle.

But let us now turn to the third part, which can be modeled on the first two, but needs a few adjustments. First, we must be able to match POS tags ("<[wc]·(. . .|. . .-. . .)>"), which are followed by words or punctuation marks ("[^<]*?"). Second, we do not want the tag-word sequences to match intervening verbs, so we modify the expression to exclude verbs: "<w·([^V]. .|[^V]. .-. . .)>[^<]*?". Finally, we need to include the information how often all of this is allowed to appear before the particle. To that end, we first group the above expression using parentheses and then add a quantifying expression in curly brackets. The final result is this:

```
>•current.matches<-exact.matches("<w•(VV.|VV.-. . .)>[^<]*?
  (<[wc]•([^V]..|[^V]..-. . .)>[^<]*?){0,5}<w•AVP>[-•a-z]*",•
  current.sentences,•pcre=T,•case.sens=F)¶
```

If the third component of this list gets printed into a file, you get the results in <C:/_qclwr/_output files/04-2-3-3_output.txt>. Again, this search will require manual editing to identify the true hits, but the error rate is not that high.

For further study/exploration: Gries (2003a)

You should now do "Exercise box 4.3: Yet other advanced regular expressions" . . .

4.2.3.4 *Potential Cognate Object Constructions*

The next example we look at is similar to the final regular expressions example in the previous chapter. Let us assume we want to retrieve possible cases of cognate object constructions. Let us further assume, for simplicity's sake, that we will apply a very coarse approach such that we only match cases where the verb precedes the potential object and we require the similarity of the verb and the object to be located at the beginnings of the words. Also, we do not specify in detail exactly what the match must look like. Let this be our corpus file:

```
>•corpus.line<-"<w•NP0>Max•<w•NP0>Grundig<c•PUN>,•<w•CRD>81<c•PUN>,•
   <w•VVD>danced•<W•AT0>a•<w•NN1>dance•<w•AV0>yesterday•<w•PRP>in•
   <w•NP0>Baden-Baden<c•PUN>."¶
```

How would we go about retrieving cognate object constructions? Well, first we need to find a verb and we need to match and be able to remember the beginning of the verb; thus our regular expression will begin with this: "<w•V. .>([^<]+)". Of course, you could be more inclusive or restrictive regarding the tags you include: this includes all verbs (i.e. also auxiliaries and modal verbs) but excludes portmanteau tags.

Then, there may be some intervening material (".*?"), and we need a noun tag ("<w•N. .>"). Finally, the characters after the noun tag must be the ones that were found at the beginning of the word immediately following the verb tag: "\\1". Thus, we write:

```
>•(current.matches<-exact.matches("<w•V..>([^<]+).*?<w•N..>\\1",•
   corpus.line,•pcre=T,•case.sens=F))[[1]]¶
[1]•"<w•VVD>danced•<W•AT0>a•<w•NN1>dance"
```

This looks good. However, there is one potential downside that you might want to take care of. The "+" in the regular expression matches one or more. However, this may be a bit too liberal since it would also match "<w•VVD>developed•<w•AT0>a•<w•NN1>dance" as you can easily see: the "d" at the beginning of "developed" is sufficient for the whole expression to match. Thus, you may want to force R to only return a match if at least two (or three, or . . .) characters are matched. Before you read on, try to find a solution to this (minor) problem.

THINK

BREAK

As you probably found out quickly, the solution would involve a range such as "{2,}" instead of a "+".

> For further study/exploration: Höche (2009)

Other interesting applications of back-referencing might be to look for repetitions of words in the context of, say, psycholinguistic research on disfluencies in adult speech or particle doubling in child language acquisition (i.e. utterances such as *he picked up the book up*).

4.2.4 A Lemma-based Concordance from Files of a POS-tagged and Lemmatized (XML) Corpus

The final example in this section involves a corpus which is annotated in XML format and whose annotation comprises both POS-tags and lemmatization, the BNC Baby (cf. the Appendix for the website). This example is worth looking at because, while the BNC has been one of the most widely used corpora and the SGML format of the BNC has been the standard for many years, more and more corpus compilers are switching to the more refined XML annotation format, and in fact the BNC is now also available in the XML format. This format is similar to the SGML annotation we know so well by now, as you can see in Figure 4.10.

```
·······<u·who="PS0YA"><s·n="165"><w·type="PNP"·lemma="i">I·</w><w·type="
VVB"·lemma="think">think·</w><w·type="PNP"·lemma="it">it</w><w·type="VB
Z"·lemma="be">'s·</w><w·type="AJ0"·lemma="hard">hard·</w><w·type="NN1"·
lemma="work">work·</w><w·type="VBZ"·lemma="in">in</w><w·type="XX0"·lemm
a="n">n</w><w·type="PNP"·lemma="it">it</w><c·type="PUN">?</c></s>¶
```

Figure 4.10 Some lines of the file kb5 from the BNC Baby.

The XML format of the BNC Baby in turn is very similar to that of the most current version of the BNC World Edition in XML format, which is exemplified in Figure 4.11.

```
<s·n="10"><w·c5="AV0"·hw="anyway"·pos="ADV">Anyway·</w><w·c5="NP0"·hw="
brenda"·pos="SUBST">Brenda·</w><unclear/></s></u><u·who="D8YPS002">
<s·n="11"><w·c5="ITJ"·hw="yes"·pos="INTERJ">Yes</w><c·c5="PUN">,·</c><w
·c5="AJ0"·hw="sure"·pos="ADJ">sure</w><c·c5="PUN">.</c></s>¶
```

Figure 4.11 Some lines of the file d8y from the BNC World Edition (XML).

As you can see, much of this is familiar. There is an utterance annotation and sentence numbering much like that of the SGML version of the BNC, which means we can again use sentence numbers to identify the lines in which we will want to search for matches—note, though, that the sentence number is not always at the beginning of the line. The tags look a little different, however, and they also contain lemma information. The tag and lemma information for words and other content items is contained in elements, which altogether consist of:

- a start tag in angular brackets that contains a name that identifies the nature of the content item they surround; after the name within the start tags, you find attributes, which are name-value pairs:
  - names identify the attribute of the content item that is characterized;
  - values identify the value of the attribute of the content item that is characterized;
- the item in question;
- an end tag in angular brackets containing the name in the start tag preceded by a "/".

For example, "<w·type="VBZ"·lemma="be">'s </w>" means:

- the content item is a word: "w";
  - the content item's type—POS type—is to be in the third person singular: "VBZ";
  - the content item's lemma is "be";
- the word in question is "'s";
- an end-of-word tag.

While this annotation scheme is more complex, it of course also provides more information and we can easily process it such that we can perform our searches on the basis of many of the lines we already used above. As an example, we will write a small script to extract candidates for *-er* nominalizations from a corpus file of the BNC Baby, i.e. words that are tagged as nouns and whose lemma ends in *-er*; examples for such nominalizations include *teacher, trader, freezer, runner, fixer-upper*, etc. Let us assume we want to do that because we are interested in analyzing cases where the *-er* nominalization process does *not* result in an agentive *-er* nominal. For example, the nominalization *runner* refers to a person/thing that performs the action denoted by the verb, as does *trader* or *freezer*. However, this kind of *-er* nominalization is not the only one: *fixer-upper* is not someone or something that fixes (up)—it is, for example, a house in need of being fixed (up). We restrict our attention to one file because the focus is on how to craft the regular expression—just use a `for`-loop to search more than one file. Also, we will first do this without `exact.matches` so that you still practice `gregexpr` and `strapply` and don't limit yourself by relying on `exact.matches` too much . . .

First, we load one such file, e.g. <C:/_qclwr/_inputfiles/corp_bncb_xml.txt>, and convert everything to lower case (since the spelling of the *-er* nominalizations does not add to important information regarding the nominalization's semantics). Second, as usual, we restrict our search to the lines that contain sentence numbers:

```
>·rm(list=ls(all=T))¶
>·current.corpus.file<-tolower(scan(choose.files(),·what="char",·
  sep="\n",·quiet=T))¶
>·current.sentences<-grep("<s·n=",·current.corpus.file,·perl=T,·value=T)¶
```

Third, we must put together a regular expression. Given what we know about the XML annotation, this is what we do. On the one hand, we want to make use of the POS tags to identify words tagged as nouns; we disregard the possibility of portmanteau tags. On the other hand, we want to make use of the lemmatization because this allows us to retrieve both singular and plural forms by just looking for a lemma ending in *-er*. In addition, we actually might want to be a bit more inclusive and also include the spelling variants *-ar* and *-or* (to retrieve words like *liar* or *conveyor*). The only other thing we now need to bear in mind is that the XML tags contain double quotes, but of course we cannot just look for double quotes because R will think that the second double quote in our line will be the closing quote for the first. Thus, we need to escape double quotes. Therefore, we will define the beginning of the tag, the first name ("type"), then the value of that name "\"n. .\" ", the second name ("lemma"), the value of the second name (something that consists of several characters that are not quotes, but only as many as needed), followed by "er" and its spelling variants, followed by the closing angular bracket and the word that is described by that whole tag.

```
>•lines.with.matches<-grep("<w•type=\"n..\"•lemma=\"[^\"]+?[aeo]r\">[^<]*",•
  current.sentences,•perl=T,•value=T)¶
```

Once we have these lines, we could print it out as a concordance in much the same way as we have done before. However, let us actually exploit R's possibilities a little more. Let's say we do not really need the context of every individual usage event, but would be more interested in seeing how often each of the potential nominalizations occurs. We can use `gregexpr` to give us the start and end positions of each match and then apply `sapply` to the output of `gregexpr` to get the overall numbers of matches for each line:

```
>•matches<-gregexpr("<w•type=\"n..\"•lemma=\"[^\"]+?[aeo]r\">[^<]*",•
  lines.with.matches,•perl=T)¶
>•number.of.matches<-sapply(matches,•length)¶
```

We then use `substr`, but since we know the length of the first argument of `substr` must be as long as the number of matches, we first generate a vector that contains the lines as often as there are matches in each line and then generate vectors of all starting and stop positions for `substr`.

```
>•lines<-rep(lines.with.matches,•number.of.matches)¶
>•starts<-unlist(matches)¶
>•stops<-starts+unlist(sapply(matches,•attributes))-1¶
>•exact.string.matches<-substr(lines,•starts,•stops)¶
```

This could be it, but since we like clean outputs, we might as well remove all tags, i.e. everything between angular brackets and remove trailing spaces. Once that is done, we can immediately generate and save a frequency table of the matches to see (i) which -er nominalizations occur at all and (ii) how often they occur.

```
>•exact.string.matches<-gsub("<.*?>",•"",•exact.string.matches,•perl=T)¶
>•exact.string.matches<-gsub("•+$",•"",•exact.string.matches,•perl=T)¶
>•sort(table(exact.string.matches),•decreasing=T)¶
```

With `strapply`, shorter versions are also available. Try to formulate at least one yourself.

THINK

BREAK

```
>•exact.string.matches<-unlist(strapply(lines.with.matches,•
  "<w•type=\"n..\"•lemma=\"[^\"]+?[aeo]r\">[^<]*",•perl=T))¶
```

The above line would give you the exact string matches before cleaning up, i.e. with all tags and maybe trailing spaces. The following line would give you only the words, but still maybe trailing spaces:

```
>•exact.string.matches<-unlist(strapply(lines.with.matches,•"<w•type=\"n..\"•
   lemma=\"[^\"]+?[aeo]r\">([^<]*)",•perl=T))¶
```

As we can see, both approaches work. Thanks to the lemmatization, we even get cases where the actual word form is in the plural (e.g. *helicopters*). However, while recall is nearly perfect, we will need to do a little post-editing to increase precision by weeding out false hits such as *car, door, number, dinner, letter, corner*, and others. We can also see that a search based only on *-er* would reduce recall a little, but increase precision a lot. However, once the false hits of whichever search have been weeded out, we can go and classify our matches according to whether the *-er* nominalization is typically agentive or not.

This concludes our section on concordancing (even though you have seen that we also used frequency-list functions). Again, there are additional ways to generate concordances that we have not dealt with. For example, we have not generated character-delimited concordance output (e.g. all your concordance lines are *n* characters long), which, however, is an output format I have used only a single time in all of my time as a corpus linguist. Be that as it may, you should now be ready to explore further ways of concordancing in the case study assignments from the website and develop your own scripts serving your individual objectives. We will now turn to collocations/collocate displays.

For further study/exploration:

- Ryder (1999), Panther and Thornburg (2002), and the references quoted therein
- R has a package for handling XML data, which is called XML and might be of interest
- if you want to print out something and delete characters (especially at the end), you need not use substr, but you can also print "\b", which is the backspace character; try this out: cat("abc\n");•cat("abc\b\n")¶

4.3 Collocations

The third method to be dealt with is concerned with lexical co-occurrence, collocation displays of the type shown in Section 2.3. (If you don't recall what these look like, have a look again because that will help you to understand what follows.) Since nearly all of what we need for collocation displays has already been discussed at length, we can deal with collocations relatively quickly. For example, we will only look at collocations in one unannotated text file because we already know how to remove tags from texts, access more than one file, etc. above. Also, we will only use a simple approach to what a word is.

Our approach will be as follows. With regard to inputting data, we will first load a corpus file and define a node word (or search word) as well as a span of elements around the node word, i.e. the range of positions around the node word from which we want to obtain the collocates. As a corpus file, we use GNU Public License, <C:/_qclwr/_inputfiles/corp_gpl_long.txt>; also, we use these settings and <C:/_qclwr/_outputfiles/04-3_output-a.txt> as our output file.

```
>•corpus.file<-tolower(scan(file=choose.files(),•what="char",•sep="\n",•
  quiet=T))¶
>•search.word<-"\\bthe\\b"¶
>•span<-3;•span<-(-span:span)•#•generate•a•vector•for•all•positions¶
>•output.file<-choose.files()¶
```

In the next step, we will have to split up our corpus into one long vector of words (I'll provide some parameters for that below) and then find out where in that vector our search word occurs. Then, we must access all words and their frequencies within the span around the node word. Since there will be differently many words in these positions, we will store the collocates at each span position and their frequencies (which we get using `table`) in different vectors in a list, because this is a data structure which, unlike data frames, can include differently long vectors. And then, when we have all that information in a list, we will loop through the list to output a table.

Before we begin to retrieve the collocates of "the", we now prepare our corpus file by transforming it into one long vector of words. These are the first seven lines of the corpus file.

```
                ·····GNU·GENERAL·PUBLIC·LICENSE¶
                ·······Version·2,·June·1991¶
¶
·Copyright·(C)·1989,·1991·Free·Software·Foundation,·Inc.¶
·51·Franklin·Street,·Fifth·Floor,·Boston,·MA··02110-1301··USA¶
·Everyone·is·permitted·to·copy·and·distribute·verbatim·copies¶
·of·this·license·document,·but·changing·it·is·not·allowed.¶
```

Figure 4.12 The first seven lines of <C:/_qclwr/_inputfiles/corp_gpl_long.txt>.

Now, let us briefly consider how we want to define the notion of word here. Let's assume we do not want to split up years (e.g. "1989") and the postal zip (i.e. "02110-1301"), but we do want to separate "'s" from the noun to which it is attached—i.e. "foundation's" would turn out to be two words. Let's assume we also want to keep hyphenated words untouched (i.e. "machine-readable" remains one word), but we want to separate words interrupted by dashes, which are represented as two hyphens in this file (recall the above word of caution: know your corpora!).

First, we put spaces around everything that is not a hyphen, a letter, a number, or already a space; as before, the hyphen is put first in the character class to make sure it is not understood as a range operator. This way, the parts separated by hyphens will still be together. Then, we clean up the file in terms of deleting leading and trailing whitespace:

```
>•cleaned.corpus.file.1<-gsub("([^-a-z0-9\\s])",•"•\\1•",•corpus.file,•
  perl=T)¶
>•cleaned.corpus.file.2<-gsub("(^\\s+|\\s+$)",•"",•cleaned.corpus.file.1,•
  perl=T)¶
```

Finally, we transform the whole corpus into one long character vector. To that end and in order to stick to our requirements from above, we split up the vector cleaned.corpus.file.2 at each occurrence of something that cannot be part of a word as we defined it or is a dash. To check whether this works—and it does—we then look at the following parts of the vector.

```
>•corpus.words.vector<-unlist(strsplit(cleaned.corpus.file.2,•
  "([^-a-z0-9]+|--)"))¶
>•corpus.words.vector[100:107]•#•to•see•"foundation"•"s"¶
>•corpus.words.vector[1218:1226]•#•to•see•"machine-readable"¶
```

Now we can determine where the occurrences of our search word are:

```
>•positions.of.matches<-grep(search.word,•corpus.words.vector,•perl=T)¶
>•head(positions.of.matches)¶
[1]••46••65••85•101•127•167
```

If there are matches in our file, we will proceed as follows. We will first generate an empty list. This list will contain as many elements as we have span positions—i.e. 7 in the present example—and each element will contain all the collocates at the respective span position.

Second, we go through each element of the vector span and use the value of each element to go from the position of the match to the position of the collocate at the desired span position. For example, we know the first match is at position 46. Thus, given a span of 3, the collocates we need to access for this first match are at the positions 43 (46−3), 44 (46−2), 45 (46−1), 47 (46+1), 48 (46+2), 49 (46+3), and of course the same applies to all other matches, too. Note also how we make use of the vector-based processing here. (I hope you notice that we discussed similar approaches above.)

```
>•results<-list()¶
>•for•(i•in•1:length(span))•{•#•access•each•span•position¶
+••••collocate.positions<-positions.of.matches+span[i]¶
```

Note: i takes on values from 1 to 7, and these values are used in span[i] to get at the span positions −3, −2, −1, 0, 1, 2, 3. Thus, if you looked at the beginning of this vector on the first iteration to see how it compares to the beginning of positions.of.matches, this is what you would see: 43, 62, 82, 98, 124, 164, . . .

As above in Section 4.2.1, there may be occasions where the first occurrence of the search word is so close to the beginning or the end of the whole corpus that the vector collocate.positions contains negative numbers or numbers greater than the corpus length. However, we are going to tabulate and the default setting of table simply discards NA, which is why we don't even have to care about those cases—we can access the words directly and immediately:

```
+••••collocates<-corpus.words.vector[collocate.positions]¶
```

Now that we have retrieved the collocates at this span position, let's determine their frequencies, sort them in descending order, and store them in the list element that is "reserved" for the current span position:

```
+••••sorted.collocates<-sort(table(collocates),•decreasing=T)¶
+••••results[[i]]<-sorted.collocates¶
+•}•#•end•of•for:•access•each•span•position¶
```

By the way, can you already guess why we sorted the table? If not, cf. below. As a result and as was mentioned above, your list results contains 7 arrays, each of which contains all collocates of the search word at one particular position such that:

- the elements are the numbers of how often a particular collocate occurred at this position; here's the beginning of the first vector ("[[1]]"), which, given a span of 3, contains the collocates three words to the left of the search word; and
- the names of the elements are the collocates at this position.

```
>•head(results[[1]])¶
••••a•••••the•••work•••••not••••••is•license
•••10•••••••9•••••••8•••••••7•••••••6•••••••6
>•head(names(results[[1]]))¶
[1]•"a"•••••••"the"•••••"work"••••"not"•••••"is"••••••"license"
```

Redundantly, the fourth vector of the list will only contain the matches of the search word, which is irrelevant here because we only had one search word, "the". Note in passing that the line results [[i]]<-sorted.collocates¶ was the reason why we did not use the simpler for-loop (with for•(i•in•1:length(span))•{¶. In this case, i would have taken on negative values and results[[i]] with i<0 would have generated an error message.

```
>•str(results)¶
List•of•7
•$•:•int•[,•1:111]•10•9•8•7•6•6•6•5•4•4•...
••..-•attr(*,•"dimnames")=List•of•1
••..•..$•collocates:•chr•[1:111]•"a"•"the"•"work"•"not"•...
•$•:•int•[,•1:106]•11•10•9•8•6•4•4•3•3•3•...
••..-•attr(*,•"dimnames")=List•of•1
••..•..$•collocates:•chr•[1:106]•"you"•"based"•"to"•"or"•...
```

```
•$•:•int•[,•1:81]•30•16•13•8•7•7•7•6•6•5•. . .
••..-•attr(*,•"dimnames")=List•of•1
••..•..$•collocates:•chr•[1:81]•"of"•"to"•"on"•"under"•. . .
•$•:•Named•int•194
••..-•attr(*,•"names")=•chr•"the"
•$•:•int•[,•1:82]•57•10•9•7•6•6•4•3•3•3•. . .
••..-•attr(*,•"dimnames")=List•of•1
••..•..$•collocates:•chr•[1:82]•"program"•"free"•"terms"•"source"•. . .
•$•:•int•[,•1:91]•27•14•12•9•8•7•6•4•4•4•. . .
••..-•attr(*,•"dimnames")=List•of•1
••..•..$•collocates:•chr•[1:91]•"of"•"or"•"is"•"software"•. . .
•$•:•int•[,•1:100]•12•11•10•9•8•8•5•5•5•4•. . .
••..-•attr(*,•"dimnames")=List•of•1
••..•..$•collocates:•chr•[1:100]•"this"•"of"•"to"•"the"•. . .
```

This output already tells you something of general interest. For every list element, the line beginning with "$·collocates:" gives you the number of different collocates observed at the corresponding span position. This is how you could get that more directly:

```
>•(lengths<-sapply(results,•length))¶
[1]•111•106••81•••1••82••91•100
```

The further away from the search word you look, the more diverse the environment becomes, something that you find for most words. At the same time, the smaller the number of collocates at a particular position, the less variation there is and the more interesting it might be to explore that position.

> For further study/exploration: Mason's (1997, 1999) work on lexical gravity for a similar and very interesting way of exploring search words' most relevant span positions, which is much more refined, but easily implementable in R

Finally, you must output the collocates and their frequencies. You will do this sequentially, listing the most frequent collocates at each position at the top. First, however, you print a column header into your output file. While this could be done with a loop (cf. the code file), a vectorized line is shorter and more elegant:

```
>•cat(paste(rep(c("W_",•"F_"),•length(span)),•rep(span,•each=2),•sep=""),•
  "\n",•sep="\t",•file=output.file)¶
```

Now to the output, which is a little tricky. We will discuss three different ways to generate very similar outputs. All three ways work, but:

- the first way is clumsy but quick and produces a good output;
- the second is really short, elegant, slightly slower, but produces a slightly inferior output;
- the third is intermediately long, intermediately elegant, takes longest, and produces the best output.[8]

The first approach is this. We know we have to print out maximally 111 lines of collocates because that is the maximum number of collocates found in the data (recall lengths for span position −3). This also means that we have to access maximally 111 positions and elements in each of the parts that make up results. We also know that we have seven positions to access. Thus, we begin with two for-loops:

```
>•for•(k•in•1:max(lengths))•{¶
+••••for•(l•in•1:length(span))•{¶
```

Now we access the position of each collocate defined by the counter k in each part of results (as defined by the counter l). This means, at the first iteration we access the first collocates in the seven parts of results. These are the most frequent words in the seven span positions (because the table is already sorted). How do we access these elements? Simple: we could use subsetting by writing results[[l]][k]¶. However, we must be prepared for cases where there is no collocate. For example, we have seen that at span position −3 there are 111 collocates, but at span position −2, there are just 106. Thus, we test whether results[[l]][k] actually accesses something. If that is the case, results[[l]][k] will not produce NA. Thus, we test whether we get NA at that element. If we do get NA, we just print two tab stops to preserve the structural identity of the table. If we do not get NA, we print the name of that element—which is the word—and the element itself—which is its frequency and complete the inner for-loop:

```
+•••••••if•(is.na(results[[l]][k])==F)•{¶
+••••••••••cat(names(results[[l]][k]),•"\t",•results[[l]][k],•"\t",•sep="",•
    file=output.file,•append=T)¶
+•••••••}•else•{¶
+••••••••••cat("\t\t",•file=output.file,•append=T)¶
+•••••••}¶
+••••}¶
```

Then, we must not forget the line break after every row, and then we complete the outer for-loop. The result file can now be easily opened into a spreadsheet software for further processing and evaluation:

```
+••••cat("\n",•file=output.file,•append=T)¶
+•}¶
```

The second way to generate the output is much, much shorter. In the spirit of Section 3.6.3, it does away with the extensive looping construction of the first approach and uses functions from the apply family. We begin again after the header has been printed into the output file (this time

<C:/_qclwr/_outputfiles/04-3_output-b.txt>) and start with a for-loop that allows us to access each of these positions:

```
>•for•(k•in•1:max(lengths))•{¶
```

For the next step, you must recall how to access parts of lists. We saw above that the same positional elements in different parts of a list can be accessed using sapply. In this case, we could access all the first elements in all vectors of the list by writing sapply(results,•"[",•1)¶. (Of course, we would need to use the looping index k below.) This retrieves the element (the frequency), and would also output the name of the element (the word). Since we need the names separately, we could just add names to the last line:

```
>•sapply(results,•"[",•1)•#•gives•the•frequencies¶
••••••a••••••you••••••of••••the•program••••••of••••this
•••••10••••••11••••••30••••194••••••57••••••27••••••12
>•names(sapply(results,•"[",•1))•#•gives•the•words¶
[1]•"a"••••••••"you"••••••"of"••••••"the"••••••"program"•"of"••••••"this"
```

We can just print them out next to each other because that would separate the collocates from their frequency rather than combining them. However, since we have two equally long structures (seven frequencies and seven names), we could paste them using tab stops as separators because these will help when importing the output file into a spreadsheet software.

```
>•paste(names(sapply(results,•"[",•1)),•sapply(results,•"[",•1),•sep="\t")¶
[1]•"a\t10"••••••••"you\t11"••••••"of\t30"
[4]•"the\t194"••••"program\t57"•"of\t27"
[7]•"this\t12"
```

Now we are nearly done. We only print out this vector with cat—now of course using k and not 1— and choose tab stops as separators for the same reason that we used tab stops as a separator with paste; don't forget to also print a line break so that the next sequence of list elements gets a new line. Our only line in the loop would be this:

```
+••••cat(paste(names(sapply(results,•"[",•k)),•sapply(results,•"[",•k),•
  sep="\t"),•"\n",•sep="\t",•file=output.file,•append=T)¶
+•}¶
```

Extremely short and elegant, isn't it? There is a small downside, though. Since the vectors in results are differently long (cf. above), the further you get, the more NA's you get, and these will be printed into the output. While this does not invalidate the results, it is just not … nice. However, with a little more code we can rectify even this. Accordingly, we turn to the third way, again beginning after the header has been printed into the output file (this time <C:/_qclwr/_outputfiles/04 -3_output-c.txt>) with a for-loop that allows us to access each of these positions:

```
>•for•(k•in•1:max(lengths))•{¶
```

Next, we paste the required elements together, but instead of directly printing it, we assign it to a vector so that we can change it before we output it. We then use a regular expression to delete the cases of NA. In order to make sure that we don't accidentally delete data where NA was a word in the corpus, we use a very explicit search pattern that includes the tab stops and replaces matches with just one tab stop.

```
+••••output.string.1<-paste(names(sapply(results,•"[",•k)),•
  sapply(results,•"[",•k),•sep="\t")¶
+••••output.string.2<-gsub("NA\tNA",•"\t",•output.string.1,•perl=T)¶
```

This was the intermediate step we would have needed above to get rid of the NAs. Now what remains is the actual output into the file before we again complete the loop:

```
+••••cat(output.string.2,•"\n",•sep="\t",•file=output.file,•append=T)¶
+•}¶
```

This discussion may have seemed a little complex. I hope, however, it has shown that there are usually many different ways to achieve one and the same goal. More importantly, I hope that entertaining these different possibilities has also raised your awareness for the ways in which you can handle tasks with R. The next section will be concerned with how to use a very differently structured kind of corpus.

> **You should now do "Exercise box 4.4: Collocations"** . . .

4.4 Excursus 1: Processing Multi-tiered Corpora

This section will be concerned with another kind of corpus file. It will deal with corpus files widely used in language acquisition research, and this example is worth dealing with because it exemplifies once more how R can provide you with data processing capabilities that go beyond what virtually all ready-made corpus software can do; just as with many scripts before, an extension of the script we will discuss in this section has also been used in actual research.

Usually, concordancing software requires that all information concerning, say, one particular sentence is on one line. In its most trivial form, this is reflected in the BNC by the fact that the expression "Wanted me to." is annotated in one line as in (11) and not in two as in (12). The latter is of course just as informative, but you may already guess that if you wanted to add further annotation to the expression "Wanted me to.", reading and processing all information in one line may quickly become unwieldy.

(11) `<w • VVD>Wanted • <w • PNP>me • <w • TOO>to<c • PUN>.`

(12) `Wanted • me • to • .`
      `<w • VVD> • <w • PNP> • <w • TOO> • <c • PUN>`

Regular concordancers such as those discussed in Wiechmann and Fuhs (2006) can usually only handle the annotation in (11), but there are corpora which are annotated as in (12), most notably perhaps those from the CHILDES database (cf. below and the Appendix for the CHILDES website), but also corpora annotated for, say, conversation-analytic or discourse-analytic purposes.[9] The following is an excerpt from the file <eve01.cha> from the Brown data of the CHILDES database (cf. Brown 1973). I only give a few lines of the header as well as the first two utterances and one later utterance. The lines with header information begin with a "@". The lines with utterances—the so-called utterance tier—begin with "*" followed by a three-character person identifier ("CHI" for the target child, "MOT" for the mother of the target child, ...) plus a colon and a whitespace character. Finally, the lines with annotation begin with "%" and a three-character annotation identifier ("mor" for "morphosyntax/lemmatization", "spa" for "speech act", "int" for "inter-action", etc.). Words and annotations are separated by spaces, and the annotation for one word consists of a POS tag, followed by "|", the lemma of the word and, if applicable, morphological annotation; cf. Figure 4.13.

```
@Font:    →   Win95:Courier:-13:1¶
@Begin  →   ¶
@Participants:  → CHI·Eve·Target_Child,·MOT·Sue·Mother,·COL·Colin¶
    →    →    →    →     Investigator,·RIC·Richard·Investigator¶
@ID: → en|brown|CHI|1;6.0|Female|||Target_Child||¶
...¶
*CHI: → more·cookie·.¶
%spa: → $IMP¶
%int: → distinctive,·loud¶
%mor: → qn|more·n|cookie·.¶
*MOT: → you·have·another·cookie·right·on·the·table·.¶
%mor: → pro|you·v|have·det|another·n|cookie·adj|right·prep|on·det|the¶
    →    →    n|table·.¶
[...]¶
*MOT: → oh·#·I·took·it·.¶
%mor: → co|oh·pro|I·v|take&PAST·pro|it·.¶
```

Figure 4.13 Excerpt of a file from the CHILDES database annotated in the chat format.

Not only is the analysis of such corpora more complicated because the information concerning one utterance is distributed across several lines, but in this case it is also more complex to load the file. If you look at the above excerpt, you will see that the compilers/annotators did not complete every utterance/annotation within only one line: the morphological annotation of the utterance "you have another cookie right on the table." stretches across two lines with the second line being indented with tab stops, so if we just read in the file as usual (scan(choose.files(),·what= "char",·sep="\n")), we cannot be sure that each element of the resulting vector is a complete utterance. Also, we cannot scan the file and specify a regular expression as a separator character (e.g. "[@%*]") because sep only allows single-byte separators. Thus, we have to be a little more creative this time and will fall back on our treatment of the ICE-GB corpus from above.

First, we read in the file <C:/_qclwr/_inputfiles/corp_chat_1.txt> as usual. Second, we collapse the whole file into one very long character string and again use a character string that does not occur in the file to preserve the information of where the line breaks had been before collapsing:

```
>•corpus.file<-scan(choose.files(),•what="char",•sep="\n")¶
>•corpus.file.2<-paste(corpus.file,•collapse="___")¶
```

We now know where every line break in the original file was, but we also know that we want to get rid of some of these, namely the ones which are followed by tab stops (cf. the line with "\t\tn| table·.¶" in Figure 4.13). Thus, we replace all occurrences of "___"—our line break marker—and at least one whitespace marker by just a single space because that is the separator used within lines:

```
>•corpus.file.3<-gsub("___\\s+",•"•",•corpus.file.2,•perl=T)¶
```

Now, in the final step, we split up the long character string at the positions where the remaining line breaks are, the ones without following tab stops that we want, to obtain a vector of corpus lines without the line breaks within utterances/annotations. In addition, we can clean up sequences of spaces that our replacement may have caused:

```
>•corpus.file.4<-unlist(strsplit(corpus.file.3,•"___",•perl=T))¶
>•corpus.file.4<-gsub("•+",•"•",corpus.file.4,•perl=T))¶
```

This vector now corresponds exactly to what we obtained in the previous examples by typing scan(choose.files(),•what="char",•sep="\n").

You should now do "Exercise box 4.5: Processing a CHILDES file 1" . . .

If we do not want to make much use of the annotation, we can proceed as before. For example, if you want to retrieve all instances of the form *phone* in the file, you write:

```
>•grep("phone",•corpus.file.4,•value=T)¶
•[1]•"*CHI:\tMommy•telephone•."
•[2]•"%mor:\tn:prop|Mommy•n|telephone•."
•[3]•"*MOT:\twell•#•go•and•get•your•telephone•."
•[4]•"%mor:\tn|well•v|go•conj:coo|and•v|get•pro:poss:det|your•n|telephone•."
•[5]•"*MOT:\tyes•#•he•gave•you•your•telephone•."
•[6]•"%mor:\tco|yes•pro|he•v|give&PAST•pro|you•pro:poss:det|your•n|telephone
   •."
•[7]•"*CHI:\tmy•telephone•."
•[8]•"%mor:\tpro:poss:det|my•n|telephone•."
•[9]•"*CHI:\ttelephone•."
[10]•"%act:\tplaying•with•toy•telephone."
[11]•"%mor:\tn|telephone•."
```

If you only want all the cases where the child uses the form *phone*, you write:

```
>•grep("^\\*CHI:.*phone",•corpus.file.4,•value=T)¶
[1]•"*CHI:\tMommy•telephone•."
[2]•"*CHI:\tmy•telephone•."
[3]•"*CHI:\ttelephone•."
```

For all instances of *that* tagged as a demonstrative ("pro:dem|that"), you enter this (note the backslashes escaping the "|"):

```
>•grep("^%mor:.*•pro:dem\\|that•",•corpus.file.4,•value=T)¶
```

This will only give you lines with annotation, not lines with actual utterances. Thus, from this output alone you will not be able to recover who used *that* as a demonstrative and what the real utterances were (since the line with the annotation only gives you the lemmatization). We will deal with a similar issue below. Finally, if you want all line numbers where the mother or the child—but no other person—uses the word *books*, you write this:

```
>•grep("^\\*(CHI|MOT):.*books",•corpus.file.4,•value=F)¶
[1]•1624•1632•1634•1652
```

In all these cases, you can use the above strategy of including more lines before and after the match if necessary (cf. Section 4.2.1). If, however, you want to use annotation, matters become more complex because you need to find a way to access information from different lines at the same time. As an example for a search relying on more than one tier, let us assume you would like to retrieve all cases where the word *telephone* or *telephones* is preceded by a possessive pronoun. While this is a task that could in principle be performed by a regular expression with many alternatives ("(my| your|his|her|its|our|their) telephone") applied to the utterance tier, we will use the annotation from the mor tier ("pro:poss:det| . . .") to exemplify the process. Another advantage of using the mor tier is that there may be cases where the child uses an expression for a possessive pronoun you could not possibly think of because its spelling deviates from the canonical adult form. Thus, the regular expression might not find these examples on the utterance tier but certainly on the mor tier where a careful annotator may have noted the creative possessive form.

There is actually a variety of different ways to handle this problem. We will use a very general one because it will allow you to apply the same logic to other cases most easily and at the same time offers the benefit of easy follow-up processing. We will convert the data from the corpus file into a spreadsheet that can be searched and processed with software such as OpenOffice.org Calc. As a first step, we reduce the corpus file to the lines relevant for searches by retrieving only the lines starting with "*" (for utterances) and "%" (for annotation), effectively deleting the header:

```
>•corpus.file.5<-grep("^[\\*%]. . .:",•corpus.file.4,•perl=T,•value=T)¶
```

Now that we have the relevant lines only, we will transform the corpus file into a list. A list will be useful here because, as we mentioned above, a list can contain several data structures, which may differ both in kind and length. First, we generate an empty list into which our results will be stored; second, we generate a counter which will allow us to keep track of where in the list we are:

```
>•corpus.file.list<-list();•counter<-0¶
```

The logic of the following is this. We want to generate a list that contains all the information in the corpus file in a structured way to store the information in vectors such that:

- there is one vector containing all the utterances;
- there is one vector containing the names of the people producing each utterance;
- there is one vector for each kind of annotation (i.e. mor, spa, int, and whatever additional annotation there may be);
- the position of each element in each of these vectors corresponds to the position of the utterance or annotation within the corpus file.

But in order for our script to be maximally flexible, we do not want to have to tell R a lot: we want R itself to recognize how many different kinds of annotation there are and to which of the utterances they apply. To that end, we will go through the corpus file line by line. Whenever we find a line that starts with a "*", we know we encounter a new utterance and we know that whatever follows until the next such line will be part of the annotation of that utterance.

```
>•for•(i•in•corpus.file.5)•{¶
```

We now extract the first character from the line, since this tells us whether this line contains a new utterance or annotation for an already processed utterance.

```
+••••first.character<-substr(corpus.file.5[i],•1,•1)¶
```

Also, we retrieve the first three characters after the "*" or the "%" because they tell us either which person produced the utterance (if the first character is a "*") or which kind of annotation we are dealing with (if the first character is a "%").

```
+••••tier.identifier•<-substr(corpus.file.5[i],•2,•4)¶
```

Finally, we retrieve the characters of the line after the colon and the tab stop until the end of the line, which is either an utterance or annotation of an utterance:

```
+••••utt.ann<-substr(corpus.file.5[i],•7,•nchar(corpus.file.5[i]))¶
```

Then, we test whether this line is a new utterance, and if it is, we will first increment our counter by 1 to start a new element in all vectors of the list:

```
+••••if•(first.character=="*")•{¶
+•••••••counter<-counter+1¶
```

Second, we will store `tier.identifier`, which must then automatically be the name of the person producing the utterance, in a vector within the list we call "names". Third, we store `utt.ann`, which we now know is an utterance, in a vector within the list that we call "utterance":

```
+•••••••corpus.file.list[["names"]][counter]<-tier.identifier¶
+•••••••corpus.file.list[["utterance"]][counter]<-utt.ann¶
```

However, if the first character is not an asterisk, we know that we are now processing annotation:

```
+••••}•else•{¶
```

Thus, we do not increment the counter by 1 because whatever annotation we now store is part of an annotation for an utterance which has already been entered into the list with a new value for counter. We only need to store the annotation, which we have already extracted into `utt.ann` into a vector within the list that has the name of the annotation.

```
+•••••••corpus.file.list[[tier.identifier]][counter]<-utt.ann¶
+••••}•#•end•of•the•conditional•expression¶
+•}•#•end•of•the•for-loop¶
```

Now, what does that buy us? Let us look at the structure of the result of this `for-loop`:

```
>•str(corpus.file.list)¶
List•of•12
•$•names••••:•chr•[1:1586]•"CHI"•"MOT"•"MOT"•"MOT"•...
•$•utterance:•chr•[1:1586]•"more•cookie•."•"you•xxx•more•cookies•?"•"how•
   about•another•graham•cracker•?"•"would•that•do•just•as•well•?"•...
•$•spa••••••:•chr•[1:1583]•"$IMP"•NA•NA•NA•...
•$•int••••••:•chr•[1:7]•"distinctive,•loud"•NA•NA•NA•...
•$•mor••••••:•chr•[1:1586]•"qn|more-n|cookie•."•"pro|you•qn|more-n|cookie-PL•
   ?"•"adv:wh|how•prep|about•det|another•n|graham•n|cracker•?"•"v:aux|would•
   det|that•v:aux|do•adj|just•prep|as•adv|well•?"•...
•$•com••••••:•chr•[1:1556]•NA•NA•NA•NA•...
•$•exp••••••:•chr•[1:1583]•NA•NA•NA•NA•...
```

```
 •$•par•••••••:•chr•[1:1375]•NA•NA•NA•NA•. . .
 •$•add•••••••:•chr•[1:974]•NA•NA•NA•NA•. . .
 •$•act•••••••:•chr•[1:1580]•NA•NA•NA•NA•. . .
 •$•sit•••••••:•chr•[1:846]•NA•NA•NA•NA•. . .
 •$•gpx•••••••:•chr•[1:1481]•NA•NA•NA•NA•. . .
```

As we can see, we now have a list which—as desired—has a vector with all utterances, a vector with all the speakers, and vectors for all kinds of annotation. For example, from the above output we can see that the first utterance ("more cookie .") is by the child (the first element of names is "CHI", that it is characterized as a imperative (spa: "$IMP"), uttered loudly and distinctively (int: "distinctive, loud"), and morphosyntactically annotated as a quantifier followed by a noun and a punctuation mark (mor: "qn|more n|cookie."), which corresponds perfectly to what we know about the makeup of the first corpus lines from the beginning of this section. You can now get at all utterances by simply calling up the part of the list with the utterances (we only look at the first 10 to save space):

```
>•corpus.file.list$utterance[1:10]¶
 •[1]•"more•cookie•."
 •[2]•"you•xxx•more•cookies•?"
 •[3]•"how•about•another•graham•cracker•?"
 •[4]•"would•that•do•just•as•well•?"
 •[5]•"here•."
 •[6]•"here•you•go•."
 •[7]•"more•cookie•."
 •[8]•"you•have•another•cookie•right•on•the•table•."
 •[9]•"more•juice•?"
 [10]•"more•juice•?"
```

Or you might want to see only the utterances by the child (again we just give the first 10):

```
>•corpus.file.list$utterance[corpus.file.list$names=="CHI"][1:10]¶
 •[1]•"more•cookie•."••••"more•cookie•."
 •[3]•"more•juice•?"•••••"Fraser•."
 •[5]•"Fraser•."•••••••••"Fraser•."
 •[7]•"Fraser•."•••••••••"yeah•."
 •[9]•"xxx•at•?•[+•bch]"•"a•fly•."
```

You can extract parts of the corpus file:

```
>•sapply(corpus.file.list,•"[",•5:10)¶
```

And, perhaps most conveniently, you can change the corpus file into a table and export it so that you process it within a spreadsheet software:

```
>•corpus.file.dataframe<-as.data.frame(sapply(corpus.file.list,•"[",•1:1586))¶
```

Note that we have to write `1:1586` at the end of the `sapply` function. As you could see from the output of `str(corpus.file.list)` above, the vectors in the list differ in length, which you know is not allowed in a data frame. With the subsetting "1:1586", we force R to use the same (maximum) number of elements in each column, and those vectors which do not have 1,586 elements (all but `names`, `utterance`, and `mor`) will be lengthened with NAs. Since this approach requires you to have a look at the list and introduce the number on your own, it is not very elegant though, since this rules out using this script for several files at the same time. Thus, how could you automatize this line more elegantly?

THINK

BREAK

As you probably figured out yourself, instead of hardcoding the number 1586 into the script, you can write this more flexible version:

```
>•max(sapply(corpus.file.list,•length))¶
```

Finally, you save the data frame into a user-defined text file such as <C:/_qclwr/_outputfiles/04 -4_chat_1-table.txt>:

```
>•write.table(corpus.file.dataframe,•choose.files(),•sep="\t",•eol="\n",•
  quote=F,•row.names=F)¶
```

You can now import this data frame into a spreadsheet software and explore the corpus with all the functions of OpenOffice.org Calc or Microsoft Excel, using filters, pivot tables, etc. (cf. <C:/_qclwr/ _outputfiles/04-4_chat_1-table.ods>).

> **You should now do "Exercise box 4.6: Processing a CHILDES file 2" ""**

Let us now tackle the issue of how to find instances of possessive pronouns followed by some form involving *phone*. An imprecise but at the same time simple way of approaching this is to first retrieve the positions of all matches of *phone* in the list's vector `utterance`. Then, you retrieve the positions of all possessive pronouns in the list's vector `mor`. Finally, you determine the positions that show up in both these searches using `intersect`:

```
>•matches.phone<-grep("phone",•corpus.file.list$utterance,•ignore.case=T,•
   value=F)•#•find•phone•-•note•you•could•use•word•boundaries,•too¶
>•matches.possessives<-grep("pro:poss:det\\|",•corpus.file.list$mor,•
   ignore.case=T,•value=F)¶
>•matches.intersection<-intersect(matches.phone,•matches.possessives)¶
>•candidates<-corpus.file.list$utterance[matches.intersection]¶
>•candidates¶
[1]•"well•#•go•and•get•your•telephone•."
[2]•"yes•#•he•gave•you•your•telephone•."
[3]•"my•telephone•."
```

In this case, this simplistic search yielded results that are perfect in terms of precision and recall. In fact, since most utterances by children are comparatively short, this crude approximation will often yield good results. However, it is also obvious that this need not be the case. If the child had said "My mommy has the telephone", this sentence would also have been matched, although it is not among the sentences we want. Given the size of most child language corpora in the CHILDES database, it would be a relatively quick and easy task to weed out such false hits.

Nevertheless, you may think it would be nice to have a more automated appropriate approach. However, this is more difficult than you may think, and it is only reasonable to point out that sometimes the nature of the corpus file(s) does not allow for a completely automated and error-free retrieval. The above strategy where a human analyst has to go over the matches and make the final decisions is probably as good as we can get. This is because there are some characteristics of the corpus file that can be dealt with in R, but which make me begin to doubt that the additional amounts of time and energy that would have to go into writing and testing code are not better spent on careful human analysis. What are these characteristics?

For example, note that in the first and second element of candidates above, a crosshatch is part of the utterance. If you look closely enough, however, you will also notice that this crosshatch is unfortunately not also part of the mor tier, which is why the elements in the utterance tier and the mor tier are not (yet) perfectly aligned. Then, closer inspection will also reveal that the same is true of the "xxx" annotation. In addition, not all words that are repeated are tagged on the mor tier as many times as they are repeated, and sometimes annotation in square brackets is not equally represented on all tiers, etc. etc. Of course, you can address some such instances relatively easily:

```
>•corpus.file.list$utterance<-gsub("•*(#|xxx|\\[.*?\\])•*",•"•",•
   corpus.file.list$utterance,•perl=T)¶
```

And of course you can find and clean up utterances which contain nothing but one expression repeated several times . . . can you?

 **THINK**

BREAK

```
>•#•find•purely•repetitive•utterances¶
>•grep("^([^•]+)•\\1\\w*$",•corpus.file.list$utterance,•perl=T,•value=T)[1:15]¶
•[1]•"man•man•."•••••••••"down•down•."•••••••"that•that•."••••••
•[4]•"that•that•."••••••••"shoe•shoe•."•••••••"quack@o•quack@o•."
•[7]•"quack@o•quack@o•."•"quack@o•quack@o•."•"stool•stool•."••••
[10]•"Mommy•Mommy•."•••••"that•that•."••••••••"soup•soup•."••••••
[13]•"down•down•."••••••"no•no•."•••••••••••"man•man•."••••••••


>•#•clean•up•purely•repetitive•utterances¶
>•corpus.file.list$utterance<-gsub("^([^•]+)•\\1\\w*$",•"\\1•",
  •corpus.file.list$utterance,•perl=T)¶
```

However, given the small number of hits, why not just look at the data yourself and make *sure* you get it all right. In my own work, I sometimes looked at thousands of hits manually to weed out false hits, etc. (and no!, contrary to what some believe, it is very often *not* enough to look at 150 examples to get an idea of the real multifacetedness and complexity of the data). That is simply part of being a corpus linguist—biologists or physicists may have to investigate hundreds or thousands of blood samples, archeologists may have to dig out many hundreds of things to find something of archeological value, well, and we have to read many matches and decide whether the match really instantiates the structure of interest . . .

Let me finally mention what you could do if you decided in favor of this more automated approach. You can now take those parts of our list you are interested in and transform them into a format you already know very well, namely a vector of lines such that each line contains tag-word sequences just as in the BNC format. As a quick grep search would show you, you can use a single "~" as our tag indicators (because it does not occur anywhere in the file):

```
>•tagged.corpus<-vector(length=length(corpus.file.list$utterance))¶
>•for•(i•in•1:length(corpus.file.list$utterance))•{¶
+••••words<-unlist(strsplit(corpus.file.list$utterance[[i]],•"•"))¶
+••••annotations<-unlist(strsplit(corpus.file.list$mor[[i]],•"•"))¶
+••••tag.word.pairs<-paste(annotations,•"~",•words,•sep="")¶
+••••tagged.corpus[i]<-paste(tag.word.pairs,•collapse="•")¶
+•}¶
```

As a result of this procedure, the vector tagged.corpus now looks like this:

```
>•tagged.corpus[38:42]¶
[1]•"n:prop|Mommy~Mommy•n|telephone~telephone•.~."
[2]•"n|well~well•v|go~go•conj:coo|and~and•v|get~get•pro:poss:det|your~your•
  n|telephone~telephone•.~."
[3]•"co|yes~yes•pro|he~he•v|give&PAST~gave•pro|you~you•
  pro:poss:det|your~your•n|telephone~telephone•.~."
```

```
[4]•"pro:wh|who~who•v|be&PRES~are•pro|you~you•v|call-PROG~calling•
  n:prop|Eve~Eve•?~?"
[5]•"pro:poss:det|my~my•n|telephone~telephone•.~."
```

Thus, if you want to find cases where a possessive pronoun is directly followed by "telephone", our regular expressions must match:

- the tag for a possessive pronoun;
- the possessive pronoun itself;
- the tag for *telephone*;
- "telephone" itself.

The tag for a possessive pronoun can be defined by the expression "pro:poss:det\\|". The possessive pronoun itself that is immediately followed by a tag and the string telephone then is ".+?~.+?~telephone". Thus, you get:

```
>•grep("pro:poss:det\\|.+?~.+?~telephone",•tagged.corpus,•perl=T,•value=T)¶
```

This produces the desired result, but you may wish to explore other options. Something else you may also wish to explore is how to change the script from above such that it can convert many different chat files into the same kind of list and data frame at the same time.

4.5 Excursus 2: Unicode

The final section in this chapter looks at non-English non-ASCII data. Just as most corpus-linguistic work has been done on English, this chapter was also rather Anglo/ASCII-centric. In this section, we are going to deal with data involving Unicode characters by briefly looking at data involving Cyrillic and Chinese characters. There is good news and bad news, however.

The good news is that nearly everything that you have read so far applies to Unicode files with Cyrillic characters, too. There are a few small changes, but nothing major. The bad news for Windows users is that in spite of some recent progress (especially R version 2.7.0) R for Windows at least is not yet as good at handling Unicode as R for MacOS X or R for Linux. There is no need for despair for Windows users, however, since many things *do* work the same way and nowadays it takes only about 20 minutes to install a version of, say, Ubuntu (<http://www.ubuntu.com>) as a dual-boot system on a PC that already runs Windows. (Even if you don't want to do that, as of Ubuntu 8.04 you can even install the opensource application Wubi on your Windows computer, which gives you access to Linux without having to change partitions or your Windows system; cf. http://wubi-installer.org/.) The following code is based on R 2.8.0 on Ubuntu 10 Intrepid Ibex, but you will see how little changes.

To load a corpus file into R, you can use the 'normal' way of reading in the file—the only difference is that, this time, choose.files() cannot be used:

```
>·rm(list=ls(all=T))¶
>·library(gsubfn)¶
>·setwd(scan(nmax=1,·what="char",·quiet=T))¶
>·source("../_scripts/exact_matches.r")¶
>·corpus.file<-scan(select.list(dir()),·what="char",·sep="\n")·#·Windows·
   users·should·add·'encoding="UTF-8"'¶
Read·23·items
>·tail(corpus.file)¶
[1]·"*·Оперативный·выезд·наших·специалистов·для·проведения·замеров."
[2]·"*·короткий·срок·изготовления·двери·на·заказ·(10·дней)."
[3]·"*·Качественная·установка."
[4]·"*·гарантия·1·год·на·все·двери."
[5]·"*·возможностъ·скидок·для·оптовых·клиентов."
[6]·"мы·работаем·для·вас!"
```

4.5.1 Frequency Lists

I will treat two ways of creating a case-sensitive frequency list of the words in this file, both of which involve the familiar strategy of splitting up the corpus file using strsplit. The first of these is basically no different from how we did it above, which is why I reduce most of the exposition to just the code, which you will understand without problems. To make it a little more interesting, however, we will first determine which characters are in the file and which of these we might want to use for strsplit, so, how do you find out all characters that are used in the corpus file?

THINK

BREAK

You need to remember that strsplit splits up (elements of) a vector character-wise if no splitting character is provided. Thus on Mac/Linux you can write:

```
>·table(unlist(strsplit(corpus.file,·"")))¶
```

This tells you pretty much exactly which characters to use for splitting (and which not to worry about here, such as semicolons and single quotes), and you can proceed as before in Section 4.1.1:

```
>·words.1<-unlist(strsplit(corpus.file,·"[--!:,\"\\?\\.\\*\\(\\)\\s\\d]+",·
   perl=T))·#·split·up·into·words¶
>·words.1<-words.1[nchar(words.1)>0]·#·delete·empty·character·strings¶
>·sorted.freq.list.1<-sort(table(words.1),·decreasing=T)·#·create·sorted·
   frequency·list¶
>·head(sorted.freq.list.1,·10)·#·look·at·result¶
words.1¶
```

```
•••••••и•••••••в•••••для•DAMINNAD•••••••к
••••••10••••••••6••••••••5••••••••4••••••••4
••••белья•••••••в••••двери•••••••на••нижнего
••••••••3••••••••3••••••••3••••••••3••••••••3
```

By the way, note that you can also apply `tolower` and `toupper` in the familiar way:

```
>•tolower(words.1);•toupper(words.1)•#•this•works,•too¶
```

One difference to the above treatment of ASCII data is that in an US-English locale, you cannot use characters such as "\\b" or "\\W" with `perl=T` here because what is a word character depends on the locale and the Cyrillic characters simply are not part of the ASCII character range [a-zA-Z]. However, there is a nice alternative worth knowing: for Cyrillic, for example, you can define your own character ranges using Unicode. In R, each Unicode character can be accessed individually by the sequence "\u" followed by a four-digit hexadecimal number.[10] This has two important consequences: first, once you know a Unicode character's number, you can use that number to call the character, which is particularly handy if, for example, you want to define a search expression in Cyrillic characters but don't have a Cyrillic keyboard and don't want to fiddle around with system keyboard settings, etc:

```
>•"\u0401"•#•Ё:•character•in•Unicode¶
[1]•Ё
>•"\u0410"•#•a:•first Russian•Cyrillic•character•in•Unicode¶
[1]•а
>•"\u044F"•#•я:•last•Russian•Cyrillic•character•in•Unicode¶
[1]•я
>•"\u0451"•#•Ё:•character•in•Unicode¶
[1]•ё
```

Thus, you can look up the characters for the Russian word meaning "for" from the Unicode website or sites at Wikipedia and then 'construct' it at the console:

```
>•(x<-"\u0434\u043B\u044F")¶
[1]•"для"
```

Second, when the characters of the alphabet you are concerned with come in ranges, you can easily define a character class using Unicode hexadecimal ranges. For instance, the small and capital characters of the Russian Cyrillic alphabet can be defined as follows:

```
> •(russ.char.capit<-"[\u0410-\u042F\u0451]")•#•all•capital•Russian•
  Cyrillic•characters¶
> •(russ.char.small<-"[\u0430-\u044F\u0401]")•#•all•small•Russian•
  Cyrillic•characters¶
```

Note how the square brackets that you need to make this a character class are already part of the character strings. This of course then also means that you can easily define non-word characters in nearly the same way:

```
> •(russ.char.yes<-"[\u0401\u0410-\u044F\u0451]")•#•all•Russian•
  Cyrillic•characters¶
> •(russ.char.no<-"[^\u0401\u0410-\u044F\u0451]")•#•all•other•characters¶
```

Thus, the alternative and maybe(!) more general way to create the above kind of frequency list is this:

```
> •words.2<-unlist(strsplit(corpus.file,•paste(russ.char.no,•"+",•
  sep=""),•perl=T))•#•split•up•into•words¶
> •words.2<-words.2[nchar(words.2)>0]•#•delete•empty•character•strings¶
> •sorted.freq.list.2<-sort(table(words.2),•decreasing=T)•#•create•sorted•
  frequency•list¶
> •head(sorted.freq.list.2,•10)•#•look•at•result¶
words.2
•••••••и•••••••в•••••для•••••••к•••белья
••••••10•••••••6••••••••5•••••••4•••••••3
•••••••в•••двери•••••••на•нижнего•••••все
•••••••3•••••••3••••••••3•••••••3•••••••2
```

If you compare this output to the one above, you should learn another important lesson: the word "DAMINNAD" shows up four times in the first list, but not in the second. This is of course because the characters that make up this word are not Cyrillic characters as defined by the Unicode character ranges, which is why they get used by `strsplit` and, thus, effectively get deleted. Also, note that just because the "A" in "DAMINNAD" and the ASCII "A" look the same to the human eye does not mean that they are the same Unicode character. For example, what may look like ASCII spaces or commas in the file <C:/_qclwr_inputfiles/corp_utf8_mandarin.txt> are not ASCII characters! Thus, you must be very careful about how to define you splitting characters.

4.5.2 Concordancing

Concordancing with Unicode data is also nearly the same as with ASCII data; the main difference being again that, with `perl=T`, you cannot use, say, "\\b". Let us begin with a simple example in which we look for a Russian word meaning "but", which is "но". You first define it as a search word (and often it is useful to immediately put parentheses around it that allow you to use back-referencing later when, for instance, you want to put tab stops around the match):

```
>•search.word<-"(но)"•#•or•(search.word<-"(\u043d\u043e)")¶
```

The simplest way to look for this is of course to just use the above search.word as the search expression with grep. However, if you look for the search expression you will find many examples where "но" is part of adjectival inflection (e.g. "ное" or "ного") or of the derivational morpheme "ность":

```
>•exact.matches(search.word,•corpus.file,•pcre=T,•case.sens=T)[[3]][6:8]¶
[1]•"Line:•3\tторговый•дом••эдера•-•компания,•развивающая•стратегические•
  от\tно\tшения•с•партнерами,•находя•индивидуальный•подход•и•учитывая•не•
  только•свои•интересы,•но•и•;интересы•каждого•из•партнеров,•для•
  Обеспечения•взаимовыгодного•сотрудничества."
[2]•"Line:•3\tторговый•дом••эдера•-•компания,•развивающая•стратегические•
  отношения•с•партнерами,•находя•индивидуальный•подход•и•учитывая•не•
  только•свои•интересы,•\tно\t•и•интересы•каждого•из•партнеров,•для•
  Обеспечения•взаимовыгодного•сотрудничества."
[3]•"Line:•3\tторговый•дом••Здера•-•компания,•развивающая•стратегические•
  Отношения•с•партнерами,•находя•индивидуальный•подход•и•учитывая•не•
  только•свои•интересы,•но•и•интересы•каждого•из•партнеров,•для•
  Обеспечения•взаимовыгод\tно\tго•сотрудничества."
```

To get rid of false hits, you might decide to find "но" only when it is used with non-word characters around it and define a search expression such this one:

```
>•(search.expression<-paste(russ.char.no,•search.word,•russ.char.no,•
  sep=""))¶
[1]•"[^Ёа-яЁ](но)[^Ёа-яЁ]"
```

But you already know that would cause problems with matches that are at the beginnings or ends of lines because then there are no preceding or subsequent chararacters respectively. You therefore have two choices: either you use define a better word boundary or you use negative lookaround. I will illustrate both options using the function exact.matches from above.

As for a better word boundary, you can add beginnings and ends of character strings to the non-Cyrillic-character range you defined earlier . . .:

```
>•(word.boundary<-paste("(^|$|",•russ.char.no,•")"•sep=""))•#•a•better•
  word•boundary¶
[1]•"(^|$|[^Ёа-яё])"
```

. . . and then create and use a new search expression:

```
>•(search.expression<-paste(word.boundary,•search.word,•word.boundary,•
  sep=""))¶
[1]•"(^|$|[^Ёа-яЁ])(но)(^|$|[^Ёа-яЁ])"
>•exact.matches(search.expression,•corpus.file,•pcre=T,•case.sens=F)[[3]]¶
[1]•"Line:•3\tторговый•дом••эдера•-•компания,•развивающая•стратегические•
  отношения•с•партнерами,•находя•индивидуальный•подход•и•учитывая•не•
  только•свои•интересы,\t•но•\ти•интересы•каждого•из•партнеров,•для•
  Обеспечения•взаимовыгодного•сотрудничества."
[2]•"Line:•5\tмолодежное•нижнее•белье•Зконом•класса•DAMINNAD•(россия):•трусы,
  •маики,•комплекты.•эта•молодежная•дерзкая•серия•нижнего•белья•
  обеспечит•вам•хорошее•настроение.•в•челом,•для•DAMINNAD•характерна•
  приверженность•к•комфортным•и•удобным•в•носке•вещам,\t•но•\тесть•и•
  стремление•к•оригинальности,•экстравагантности.•ведь•DAMINNAD•-•это•
  важнейшее•средство•самовыражения,•а•молодость•-•именно•то•время,•когда•
  позволено•все•и•когда•любая•одежда•тебе•к•лищу.•в•общем,•молодо•и•
  позитивно!!!"
```

As you can see, now only the one desired match in the third line is found plus another one in the fifth line.

The alternative involves negative lookaround: you look for the search word, but only if there is no character from the Cyrillic character range in front of it and no character from the Cyrillic character range after it. As you can see, the same matches are returned:

```
>•(search.expression<-paste("(?<!",•russ.char.yes,•")",•search.word,•
  "(?!",•russ.char.yes,•")",•sep=""))¶
>•exact.matches(search.expression,•corpus.file,•pcre=T,•case.sens=F)[[3]]¶
```

However, since none of these matches actually involves a match at the beginning or end of a string, let's just make sure that this works by looking for the Russian translation of *in*, "в":

```
>•search.word<-"(в)"•#•or•(search.word<-"(\u0432)")¶
```

If you define the search expressions just as above and then perform both searches with `exact.matches`, you will see that matches of "в" within words are not returned—as desired—but that the line-initial match in the eighth line is found:

```
>•(search.expression<-paste(word.boundary,•search.word,•word.boundary,•
  sep=""))¶
>•(search.expression<-paste("(?<!",•russ.char.yes,•")",•search.word,•
  "(?!",•russ.char.yes,•")", sep=""))¶
```

This concludes our brief excursus on Unicode character searches in this section. However, since this is an important issue, I would like to encourage you not only to get more familiar with this kind of encoding (cf. the references mentioned in Section 2.1 above), but also to do Exercise box 4.7 now.

You should now do "Exercise box 4.7: Unicode" . . .

5

Some Statistics for Corpus Linguistics

[I]t seems to me that future research should deal with frequencies in a much more empirically sound and statistical professional way. [. . .] In the long run, linguistic theorising without a firm grip on statistical methods will lead corpus linguistics into a dead-end street.

(Mukherjee 2007: 140f.)

Let me first briefly say something about the role I think statistical analysis should play in corpus linguistics. I actually believe that statistical analysis is an indispensable ingredient of corpus-linguistic work even though, unfortunately, it has not figured as prominently in corpus linguistics as one would like it to: even now, there is still much work that contents itself with reporting raw (co-) occurrence frequencies and neither employs the most basic tests of statistical significance nor reports standard measures of effect size. Even worse, even otherwise excellent textbooks exhibit quite a few errors when it comes to statistics. However, since corpus analyses invoke frequencies and distributional patterns, one should evaluate them with the methods that are uniformly recognized as the most adequate analytic tools for frequencies and distributional patterns, which are the methods of exploratory, descriptive, and inferential statistics.

In this chapter, you will learn about ways to handle the two kinds of dependent variables that corpus analyses usually invoke: frequency data and interval/ratio-scaled data (such as lengths of words). For each kind of data, I will first illustrate a way of representing the results graphically and then introduce a statistical analysis required for null-hypothesis significance testing.

Since this book cannot and does not attempt to provide a fully fledged *and* practical introduction to corpus-linguistic statistics—in fact, I am not aware of such a textbook—I will introduce only some ways in which R can be applied to the evaluation of several frequent corpus-linguistic research designs. These statistical techniques will therefore only provide a start for simpler kinds of analysis—I can only scratch the surface. For some of the more complex methods, I will provide the R functions you would need for them, but I strongly recommend that you first do some serious reading that provides more statistical background than I can offer here. Some of my favorite introductions to statistics are Crawley (2002, 2005, 2007), using S-Plus and R. Baayen (2008) is an excellent though very demanding introduction to statistics for linguists, and Johnson (2008) is also very good but less challenging. Other very good introductory references that introduce some statistics with R in more detail are Dalgaard (2002) and Good, P. (2005). Maindonald and Braun (2003) is a good though somewhat more demanding textbook reference to more advanced techniques. Lastly, Gries (to appear) covers many of the methods mentioned here and more and how to apply them with R in more detail and not too technically.

5.1 Introduction to Statistical Thinking

Before we begin to actually explore ways of how corpus data can be analyzed statistically with R, we need to cover a few basic statistical concepts. The first concept is that of a variable. A variable is a symbol of a set of values or levels characterizing an entity figuring in an investigation; in the remainder of this chapter, variables will be printed in small caps. For example, if you investigate the lengths of subjects (by counting their lengths in words), then we can say the subject of *The man went to work* gets a value of 2 for the variable LENGTH. There are two ways in which variables must be distinguished: in terms of the role they play in an analysis and in terms of their information value, or level of measurement.

5.1.1 Variables and their Roles in an Analysis

You will need to distinguish between dependent and independent variables. Dependent variables are the variables whose behavior/distribution you investigate or wish to explain; independent variables are the variables whose behavior/distribution you think correlates with what happens in the dependent variables. Thus, independent variables are often, but not necessarily, the causes for what's happening with dependent variables.[1]

5.1.2 Variables and their Information Value

Here, there are four classes of variables you will need to distinguish. The variable class with the lowest information value is that of categorical variables. If two entities A and B have different values/levels on a categorical variable, this means that A and B belong to different classes of entities. However, even if the variable values for A and B are represented using numbers, the sizes of the numbers do not mean anything. For example, if one codes the two NPs *the book* and *a table* with respect to the variable DEFINITENESS, then one can code the fact that the NP *the book* is definite with a "1" and the fact that a table is indefinite with a "2", but nothing hinges on the choice of any particular number or the fact that the value representing indefinite NPs is twice as high as that of definite NPs. In fact, you could equally well code definite NPs with a "2" and indefinite NPs with a "1"; you could even code definite NPs with "34.1" and indefinite NPs with "–2": the size of the numbers just does not matter—the only thing that matters is that there are two different numbers that represent two different levels of the variable DEFINITENESS. Other examples for categorical variables (and their possible levels) are:

- phonological variables: STRESS (stressed vs. unstressed), STRESSPOSITION (stress on first syllable vs. stress elsewhere),
- syntactic variables: NP-TYPE (lexical NP vs. pronominal NP), CONSTRUCTION (V NP PP vs. V NP NP),

- semantic variables: ANIMACY (animate vs. inanimate), CONCRETENESS (concrete vs. abstract), LEXICALASPECT (activity vs. accomplishment vs. achievement vs. state),
- other examples: PARTYPREFERENCES (Republican vs. Democrats vs. Greens), TESTRESULT (pass vs. fail).

In the case of dichotomous categorical variables (also referred to as nominal or binary variables), one usually uses the numbers 0 and 1 as numerical codes. On the one hand, this has computational advantages; on the other hand, this coding often allows one to conceptualize the coding as reflecting the presence ("1") and absence ("0") of some feature. For example, one could code ANIMACY: animate as "1" and ANIMACY: inanimate as "0".

The variable class with an intermediate information value is that of ordinal values. Just like categorical variables, ordinal variables provide the information that entities coded with different values are different, but in addition they also allow for rank-ordering the entities on the basis of these values—however, they do not allow to precisely quantify the difference between the entities. One example would be the variable SYNTACTICCOMPLEXITY of NPs. It is probably fair to assume that English NPs can be ordered in terms of their complexity as follows:

(13)   pronominal NPs (e.g. *he*) < simple lexical NPs (e.g. *the book*) < non-clausally modified lexical NPs (e.g. *the green book* or *the book on the table*) < clausally modified NPs (*the book I bought last week*).

An ordinal coding of SYNTACTICCOMPLEXITY could now be this:

(14)   pronominal NPs ("1") < simple lexical NPs ("2") < non-clausally modified lexical NPs ("3") < clausally modified NPs ("4").

In this coding, higher numbers reflect higher degrees of complexity so NPs can be ranked according to their complexity. However, note that the numbers that represent different degrees of complexity do not necessarily reflect the exact differences between the different NPs. Just because a non-clausally modified NP is coded as "3" while a pronominal NP is coded as "1", this does not mean that the former is exactly three times as complex as the latter. Another linguistic example would be the IDIOMATICITY of a VP, which may be distinguished as "high (completely idiomatic)" vs. "intermediate (metaphorical)" vs. "low (fully compositional)". A final, non-linguistic example would be grades: a student with an A (4.0) is not necessarily 1⅓ times as good as a student with a B (3.0) or four times as good as a student with a D (1.0).

The next variable class to be distinguished here is that of interval variables. Interval variables provide the information of ordinal variables, but also provide meaningful information about the difference between different values. We will not treat interval variables here in much detail but rather focus on the, for our purposes, more useful class of ratio variables (interval and ratio variables together are sometimes also referred to as continuous variables). Ratio variables provide the information of interval variables, but in addition they also have a clear definition of 0, an absolute, non-arbitrary zero point. For example, the syllabic length of an NP is a ratio variable because it makes sense to say that an NP with four syllables is twice as long as one with two syllables and four times as long as one with just one syllable. Other examples for ratio variables would be pitch frequencies in Hz, word frequencies in a corpus, number of clauses between two successive occurrences of two ditransitives in a corpus file, the reaction time towards a stimulus in milliseconds . . .

We can now exemplify the differences between the different kinds of variables by looking at the results of a 100 m dash at the Olympics represented in Table 5.1.

Table 5.1 Fictitious results of a 100 m dash

Time	Rank	Name	StartNumber	Medal
9.86	1	Steve Davis	942235	1
9.91	2	Jimmy White	143536	1
10.01	3	Carl Lewis	346377	1
10.03	4	Ben Johnson	436426	0

Which column exemplifies which kind of variable?

THINK

BREAK

TIME is a ratio variable: the differences and ratios between different times are meaningful. RANK, by contrast, is an ordinal variable only: the runner ranked first did not need half the time of the runner ranked second although his rank is half the size of that of the second. NAME is a categorical variable, as is STARTNUMBER (because the size of the start number does not mean anything). Finally, MEDAL is a categorical variable but has been coded with "0" and "1" so that it could in principle also be understood as an ordinal variable (higher ranks meaning faster races) or a ratio variable (representing the number of medals a runner was awarded).

This classification of variables in terms of their information value will later be important to choose the right statistical technique for the evaluation of data because not every statistical technique can be applied to every kind of variable.

5.1.3 Hypotheses: Formulation and Operationalization

One of the most central notions in statistical analysis is that of a hypothesis. The term *hypothesis* is used here in a somewhat stricter sense than in the everyday sense of "assumption". Following Bortz (2005: 1–14), in what follows, hypothesis refers to:

- a statement characterizing more than a singular state of affairs;
- a statement that has at least implicitly the structure of a conditional sentence (with *if . . ., then . . .* or *the more/less . . ., the more/less . . .*);
- the hypothesis must be potentially falsifiable.

While the first criterion needs no additional explanation, the second criterion requires a little elaboration. The logic behind this criterion is that, while hypotheses typically have one of the above syntactic forms, they need not. All that is required is that the statement can be transformed into a conditional sentence. For example, the statement *On average, English subjects are shorter than English objects* does not explicitly instantiate a conditional sentence, but it can be transformed into one: *If a constituent is an English subject, it is on average shorter than a constituent that is an English object*. Or, for example, the statement *More frequent words have more senses than less frequent words* does not explicitly instantiate a conditional sentence, but it can be transformed into one: *The more frequent a word is, the more senses it has*.

The third criterion should also briefly be commented on since it must be understood in two ways. On the one hand, it means that there must be conceivable states of affairs that show the hypothesis to be false. Thus, *Drinking alcohol may influence the reaction times of subjects in an experiment* is not a hypothesis in this sense because the *may influence* basically means 'may influence or may not', so every result that may occur in an experiment—an influence or a lack of it—is compatible with the hypothesis. On the other hand, there are potentially falsifiable statements which qualify as hypotheses, but which cannot be tested nowadays for, for instance, ethical reasons. A particularly infamous example is the question of which language children speak if they are raised without any linguistic input, or the hypothesis that, when children are raised without any linguistic input, then they will speak Greek. This hypothesis is of course a hypothesis, meeting all criteria from above, but while this hypothesis appears to have actually been tested (cf. Steinberg 1993: Section 3.1), we obviously cannot test it again, given the blatant violation of ethical standards.

Above, hypotheses were characterized as invoking the notion of conditionality. This characterization anticipates the first central distinction of variables, that between independent and dependent variables. Simplistically speaking, independent variables are variables that are mentioned in the *if* clause or the first *the more/less* clause, while dependent variables are variables mentioned in the *then*-clause or the second *the more/less* clause. It is important to understand, however, that there are two parameters according to which hypotheses are classified.

The first parameter refers to the contrast between the so-called alternative hypothesis (abbreviated as H_1) and the so-called null hypothesis (abbreviated as H_0). Usually, the former is a hypothesis that,

- if you look at just one variable, states that the values/levels of the dependent variable do not follow a random distribution of the investigated entities (or some other expected distribution such as the normal distribution);
- if you look at dependent and independent variables, states that the values/levels of the dependent variable vary as a function of the values/levels of the independent variable(s).

For example, *On average, English subjects are shorter than English objects* would be an alternative hypothesis because it states that some non-trivial part of the variation you find in the lengths of XPs is due to fact that the XPs instantiate different grammatical relations, namely subjects and direct objects. The null hypothesis, by contrast, is the logical counterpart of the alternative hypothesis and usually states that,

- if you look at just one dependent variable, it states that the values/levels of the dependent variable are randomly distributed or correspond to some specified distribution (such as the normal or the binomial distribution);
- if you look at dependent and independent variables, it states that the values/levels of the dependent variable do not vary as a function of the values/levels of the independent variable(s).

You can often form a null hypothesis by inserting *not* into the alternative hypothesis (that is, *On average, English subjects are* not *shorter than English objects*), but typically the null hypothesis states there is no effect (that is, *On average, English subjects and English objects do not differ in their lengths* or *On average, English subjects and English objects are equally long*).

The second parameter is concerned with the language of the hypothesis, so to speak. Both the alternative hypothesis and the null hypothesis come in two forms. The first of these is the one you have already seen, a text form in natural human language. The other form is the statistical form in mathematical language, which brings me to the issue of operationalization.

Operationalization refers to the process of deciding how the variables in your hypotheses shall be investigated. In other words, you answer two interrelated questions. First, "what will I perceive when

I perform our study and observe the values/levels of the variables involved?" Second, "which mathematical concept will I use—counts/frequencies, averages, dispersions, or correlations?" This is necessary because, often, the formulation of the hypotheses in text form leaves open what exactly will be counted, measured, etc. In the above case, for example, it is not yet clear how you will know whether a particular subject is longer than a particular direct object.

(15)   The younger bachelors ate the nice little parrot.

With regard to the former question, this sentence can support either the alternative hypothesis or the null hypothesis or neither of the two, depending on how you operationalize the variable LENGTH. If you operationalize LENGTH as "number of morphemes", the subject gets a value of 5 (*The*, *young*, comparative *-er*, *bachelor*, plural *s*) and the direct object gets a value of 4 (*the*, *nice*, *little*, *parrot*). On the other hand, if you operationalize LENGTH as number of words, the subject and the object get values of 3 and 4 respectively. Finally, if you operationalize LENGTH as "number of letters (without spaces)", both subject and object get a value of 19.

With regard to the latter question, you also must decide on a mathematical concept. For example, if you investigate, say, 99 sentences and choose to operationalize lengths as numbers of morphemes, your hypotheses could involve averages:

H_0:   The average length of the subjects in morphemes is the same as the average length of the direct objects in morphemes; mean length of subjects = mean length of direct objects.

H_1:   The average length of the subjects in morphemes is different from the average length of the direct objects in morphemes; mean length of subjects ≠ mean length of direct objects.

. . . or frequencies:

H_0:   The number of cases where the subject is longer (in morphemes) than the direct object is the same as the number of cases where the subject is shorter (in morphemes) than the direct object.

H_0:   The number of cases where the subject is longer (in morphemes) than the direct object is different from the number of cases where the subject is shorter (in morphemes) than the direct object.

. . . or even correlations, which I will discuss below. Thus, you must exercise care in operationalizing your variables: your operationalization of variables determines the validity of the investigation, i.e. whether you measure/investigate what you intend to measure/investigate.

> **You should now do "Exercise box 5.1: How would you operationalize these variables? Why?" . . . and "Exercise box 5.2: Formulate alternative and null hypotheses in text form and in statistical form" . . .**

While the discussion so far has focused on the situation where we have maximally two variables, a dependent and an independent one, of course life is usually not that simple: there is virtually always more than one determinant for some phenomenon. While we will not deal with statistical techniques to deal with several independent variables here, it is important for you at least to understand how such phenomena would be approached (cf. Baayen 2008 and Gries to appear for some examples and discussion of the relevant techniques). We have been talking about the alternative hypothesis that, on average, English subjects are shorter than English objects, which involves one dependent variable (LENGTH) and one independent variable (GRAMMATICAL RELATION: subject vs. object). For the sake of the argument, let us imagine you now suspect that the lengths of some constituents does not only vary as a function of their being subjects and objects, but also in terms of whether they occur in main clauses or subordinate clauses. Technically speaking, you are introducing a second independent variable, CLAUSE TYPE, with two levels, main clause and subordinate clause. Let us also assume you do a pilot study—without having a hypothesis yet—in which you look at 120 subjects and objects (from a corpus) with the following numbers of cases:

Table 5.2 Fictitious data set for a study on constituent lengths

	Grammatical relation: subject	Grammatical relation: object	Totals
CLAUSE TYPE: main	30	30	60
CLAUSE TYPE: subordinate	30	30	60
Totals	60	60	120

You then count for each of the 120 cases the length of the subject or object in syllables and compute the mean lengths for all four conditions, subjects in main clauses, objects in main clauses, subjects in subordinate clauses, and objects in subordinate clauses. Let us assume these are the main effects you find, where a main effect is the effect of one variable in isolation:

- constituents that are subjects are shorter than constituents that are objects;
- constituents in main clauses are shorter than in subordinate clauses.

There are now two possibilities for what these results can look like (or three, depending on how you want to look at it). On the one hand, the two independent variables may work together additively. That means the simultaneous combination of levels of independent variables has the effect that we would expect on the basis of their individual effects. In this case, you would therefore expect that the shortest constituents are subjects in main clauses while the longest constituents are objects in subordinate clauses. One such result is represented in a so-called interaction plot in Figure 5.1.

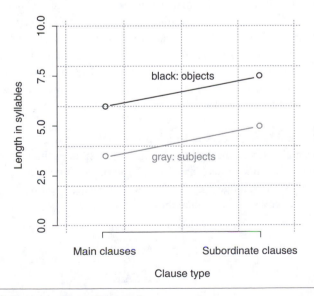

Figure 5.1 Interaction plot for GRAMMATICAL RELATION × CLAUSE TYPE 1.

However, the independent variables may not work together additively, but interact. Two variables are said to interact if their joint influence on the dependent variable is not additive, i.e. if their joint effect cannot be inferred on the basis of their individual effects alone. In our example: if direct objects are longer than subjects and if elements in subordinate clauses are longer than elements in main clauses, but you also find that objects in subordinate clauses are in fact very short, then this is called an interaction of GRAMMATICAL RELATION and CLAUSE TYPE. One such scenario is represented in Figure 5.2.

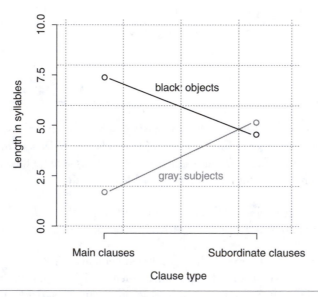

Figure 5.2 Interaction plot for GRAMMATICAL RELATION × CLAUSE TYPE 2.

As you can see, the mean length of subjects is still smaller than that of the objects. Also, the mean length of main clause constituents is still smaller than that of the subordinate clause constituents, but the combinations of these levels do not result in the same predictable effects as before in Figure 5.1. In other words, while objects are in general longer than subjects—the main effect—not all objects are: the objects in subordinate clauses are not only shorter than objects in main clauses, but in fact also shorter than subjects in subordinate clauses, but longer than subjects in main clauses. Thus, if there is an interaction of this kind—which is often easy to recognize because of the crossing lines in such an interaction plot, then you need to qualify the interpretations of the main effects.

There is yet another kind of interaction, which is represented in Figure 5.3. This may strike you as strange. Again, the main effects are the same: subjects are shorter than objects and main clause constituents are shorter than subordinate clause constituents. And, even the prediction that objects in subordinate clauses are the longest elements and subjects in main clauses are the shortest elements is borne out. So why is this an interaction?

THINK

BREAK

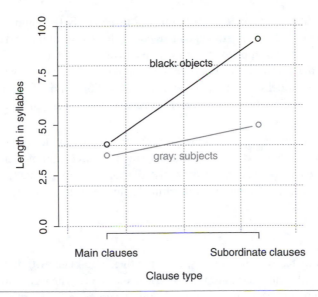

Figure 5.3 Interaction plot for GRAMMATICAL RELATION × CLAUSE TYPE 3.

It is still an interaction because while the lines in the interaction plot do not intersect, the slope of the black line is much steeper than that of the gray line. Put differently, while the difference between the means of subjects and objects in main clauses is only one syllable, it is four times larger in subordinate clauses, and because of this unexpected element, this is still an interaction.

Interactions are a very important concept, and whenever you consider or investigate designs with multiple independent variables, you must be careful about how to formulate your hypotheses—sometimes you may be more interested in an interaction than a main effect—and how to evaluate and interpret your data.

One final aspect is noteworthy here. Obviously, when you formulate hypotheses, you have to make many decisions: which variables to include, which variable levels to include, which kinds of relations between variables to consider (additive vs. interaction), etc. One important guideline in formulating hypotheses you should always adhere to is called Occam's razor. Occam's razor, named after William of Ockham (1285–1349), is the name for the following Latin principle: *entia non sunt multiplicanda praeter necessitatem*, which means 'entities should not be multiplied beyond necessity' and, generally boils down to 'try to formulate the simplest explanation or model possible'. I use *model* here in the way defined by Johnson (2008: 106), referring to a "proposed mathematical description of the data with no assumptions about the possible mechanisms that cause the data to be the way they are". More specifically, this principle—also referred to as the principle of parsimony or succinctness—includes, but is not limited to, the following guidelines:

- don't invoke more variables than needed: if you can explain the same amount of variability of a dependent variable with two models with different numbers of independent variables, choose the model with the smaller number of independent variables;
- do not invoke more variable levels than necessary: if you can explain the same amount of variability of the dependent variable with two models that have different numbers of levels of independent variables, choose the model with the smaller number of levels of independent variables; for example, if you can explain 60 percent of the variation in the data with a variable that distinguishes two levels of animacy (animate vs. inanimate), then

prefer this variable over one that distinguishes three levels (e.g. human vs. animate vs. inanimate) and can also only explain 60 percent of the variation in the data;

- choose additive relationships over relationships involving interactions: if you can explain 60 percent of the variation of the dependent variable with a model without interaction, then prefer this model over one that involves interactions and can also only explain 60 percent of the variation. (This is basically a corollary of the first item of this list.)

One motivation for this principle is of course simplicity: simpler models are easier to explain and test. Also, this approach allows you to reject the unprincipled inclusion of ever more variables and variable levels if these do not explain a sufficiently substantial share of the data. Thus, you should always bear this principle in mind when you formulate your own hypotheses.

5.1.4 Data Analysis

When you have formulated your hypotheses, and only then, you begin with the analysis of the data. For example, you might write a script to retrieve corpus data, etc. The next important point is how you store your data for the subsequent analysis. I cannot remember how often I have seen students print out long lists of data to annotate them using a paper-and-pencil approach or open long concordances in text processing software. While this may be a doable shortcut if you have small data sets, the better way of handling data is to always—and I mean *always!*—store your data in a data frame that you either edit in R or, more conveniently, in a spreadsheet software such as OpenOffice.org Calc. There are a few crucial aspects to your handling of the data, which I cannot emphasize enough. First, every row represents *one and only one* analyzed case or data point. Second, every column but the first represents either the data in question (e.g. a concordance line to be investigated) or *one and only one* variable with respect to which the data are coded. Third, the first column just gets a counter from 1 to *n* so that you can always restore the data frame to one particular (the original?) order. In corpus studies, it is often useful to also have columns for variables that you might call "source of the data"; these would contain information regarding the corpus file and the line of the file where the match was obtained, etc.

Let us assume you investigated the alternative hypothesis *Subjects are longer than objects* (operationalizing length as "length in words"). First, a think break: what is the independent variable, and what is the dependent variable?

THINK

BREAK

The independent variable is the categorical variable GRAMMATICAL RELATION, which has the levels subject and object, and the dependent variable is interval variable LENGTH. If you now formulated all four hypotheses and decided to analyze the (ridiculously small) sample in (16),

(16) a. The younger bachelors ate the nice little parrot.
      b. He was locking the door.
      c. The quick brown fox hit the lazy dog.

then your data frame should *not* look like this, an arrangement students often use:

Table 5.3 A bad data frame

Sentence	Subject	Object
The younger bachelors ate the nice little parrot.	3	4
He was locking the door.	1	2
The quick brown fox hit the lazy dog.	4	3

The data frame in Table 5.3 violates all of the above rules: first, every row contains two data points and not just one. Second, it does not have a column for every variable. Rather, leaving aside the first column, which was only included for expository purposes, it has two columns, each of which represents one *level* of the independent variable GRAMMATICAL RELATION. Before you look below, think about how you would have to reorganize Table 5.3 so that it conforms to the above rules.

THINK

BREAK

This is how your data frame should look. Now every observation has one and only one row, and every variable—independent and dependent—has its own column:

Table 5.4 A better data frame

CASE	SENTENCE	RELATION	LENGTH
1	The younger bachelors ate the nice little parrot.	subject	3
2	The younger bachelors ate the nice little parrot.	object	4
3	He was locking the door.	subject	1
4	He was locking the door.	object	2
5	The quick brown fox hit the lazy dog.	subject	4
6	The quick brown fox hit the lazy dog.	object	3

An even more precise variant may have an additional column listing only the subject or object that is coded in each row, i.e. listing *The younger bachelors* in the first row. As you can see, every data point—every subject and every object—gets its own row and is in turn described in terms of its variable levels in the two rightmost columns. This is how you should virtually always store your data. More complex statistical procedures than the ones discussed here occasionally require a slightly different layout, but this is the default arrangement. Ideally, you enter your data with a spreadsheet software (such as OpenOffice.org Calc) and then save the data (i) in the native format of that software (to preserve colors or other formatting) that would not be preserved in a text file and (ii) as a text file (for easier handling with R or other software).

5.1.5 Hypothesis (and Significance) Testing

Once you have created a data frame such as Table 5.4 containing all your data, you must evaluate the data. As a result of that evaluation, you obtain frequencies/counts, averages, dispersions, or correlations. The most essential part of the statistical approach of hypothesis testing is that, contrary to what you might expect, you do *not* try to prove that your alternative hypothesis is correct—you try to show that the null hypothesis is most likely(!) not true, and since the null hypothesis is the logical counterpart of your alternative hypothesis, this in turn lends credence to the alternative hypothesis. In other words, in most cases you wish to be able to show that the null hypothesis can *not* account for the data so that you can adopt the alternative hypothesis. You may now ask yourself, "why's that?" The answer to this question can in part be inferred from the answers to these two questions:

- How many subjects and direct objects would one have to check maximally to determine that the above alternative hypothesis H_1 is correct?
- How many subjects and direct objects would one have to check minimally to determine that the above null hypothesis H_0 is false?

THINK

BREAK

As you may have figured out yourself, the answer to the first question is infinity: strictly speaking, you can only be certain your alternative hypothesis is correct when you looked at *all* cases and found no example that does not conform to the alternative hypothesis.

The answer to the second question is "one each" because if the first example contradicted your null hypothesis, you could strictly speaking already abandon your null hypothesis since there's now at least one case where it doesn't hold.

In the social and behavioral sciences, however, one does not usually stop after the first element falsifying the null hypothesis. Rather, one collects and evaluates a data set and then determines which of the two statistical hypotheses—H_1 or H_0—is more compatible with the result. More specifically, and this is now the single most important point, if the probability p to get the obtained result when the null hypothesis H_0 is true is 0.05 (i.e. 5 percent or greater), then one cannot accept the alternative hypothesis H_1; otherwise, you are allowed to reject the null hypothesis H_0 and accept the alternative hypothesis H_1.

This may seem a little confusing because there are suddenly two probability values p: one that says how likely you are to get your result when the null hypothesis H_0 is true, another one which is mostly set to 0.05. This latter probability p is the probability not to be exceeded to still accept the alternative hypothesis H_1. It is called "significance level" and is *defined before* the analysis. The former probability p to get the obtained result when the null hypothesis H_0 is true is the probability to err when accepting H_0. It is therefore called the "probability of error" or "the p-value" and is computed on the basis of your data, i.e. *after* the analysis. But how is this probability computed and used? We will not answer this question at a particularly high level of mathematical sophistication, but since this whole complex of notions is so crucial to the understanding of statistical testing, we will still look at one example.

Let us assume you and I toss a coin 100 times. When heads is up, I get $1 from you; when tails is up, you get $1 from me. Let us assume the following hypotheses:

Text form of H_0: You and I play honestly with a fair coin; the likelihood for heads and tails for each toss is 50 percent vs. 50 percent.

Text form of H_1: I cheat to increase the likelihood of winning from the expected 50 percent to something higher.

Since this scenario cries out for an operationalization in terms of counts (of wins and losses), the following statistical hypotheses result:

Statistical form of H_0:   You and I will win the same number of times; that is, the expected frequencies of wins for both of us are $wins_I = wins_{you} = 50$.

Statistical form of H_1:   I will win more often than 50 times; that is, the expected frequencies of my wins is higher than that of your wins: $wins_I > wins_{you}$.

Now, a question: if you were my opponent, which results after 100 tosses would make you suspicious and accept the alternative hypothesis H_1 (that I am cheating)?

 THINK

BREAK

- when you lose 55 times?
- when you lose 60 times?
- when you lose 70 times?

Without probably even knowing it, you are currently doing a significance test . . . Let us make this a little more concrete and assume that you lost sixty times. Let us also assume you would only accuse me of cheating if the likelihood of getting this result or an even worse result—worse for you, that is—is smaller than 0.05. More technically, you set your significance level to 0.05. Now you are thinking, "How likely is it—i.e. what is the p-value—that I win only 40 times (the observed frequency of $wins_{you}$) when the null hypothesis H_0 was true and I should have won 50 times (the expected frequency of $wins_{you}$)?" That is, you are effectively trying to determine the p-value of the result you expected to see to compare it to your significance level.

The probability of losing 60 times or more often in 100 tosses when H_0 is true (i.e. nobody is cheating) is 0.02844397.[2] In other words, the chance of getting such an extreme result is smaller than your significance value of 0.05 so, according to your standards, you can now declare the result to be significant, adopt the alternative hypothesis H_1, and accuse me of cheating.[3] If you and I were good friends, you might have set your significance level to only 1 percent before the game because you wanted to make sure you only accuse me of cheating when there is a really exceptional result, and then you could not accuse me of cheating because 0.02844397 > 0.01.

While this example hopefully made the logic behind statistical testing clearer, it still did not say much about how exactly the p-values are computed other than note 2. Let us therefore reduce the example to more manageable proportions: we toss the coin only three times. The three left columns of Table 5.5 provide the complete result space, i.e. all possible results that you and I may achieve; each row is one possible outcome of three tosses.

Table 5.5 All possible outcomes of tossing a coin three times and their individual probabilities (when H_0 is true)

Toss 1	Toss 2	Toss 3	# heads	# tails	p_{result}
heads	heads	heads	3	0	0.125
heads	heads	tails	2	1	0.125
heads	tails	heads	2	1	0.125
heads	tails	tails	1	2	0.125
tails	heads	heads	2	1	0.125
tails	heads	tails	1	2	0.125
tails	tails	heads	1	2	0.125
tails	tails	tails	0	3	0.125

Columns 4 and 5 sum up how many heads and tails are obtained on every possible outcome. Obviously, the first result only has heads and no tails, etc. Finally, the rightmost column gives the probability of each row's result when the null hypothesis is true. As you can see, they are all identical, $p = 0.125$. There are two related ways of arriving at this probability. One would be to say, the null hypothesis says heads and tails are equally likely and the successive trials are independent. Thus, if there are altogether eight possible outcomes and each of them is equally likely, then each of them must be $\frac{1}{8} = 0.125$. The other would be to compute the probability eight times. For the first row: the probability to get heads on the first toss is 0.5. The probability to get heads on the second is again 0.5; thus the probability for two heads on the first two tosses is $0.5 \cdot 0.5 = 0.5^{\text{number of tosses}} = 0.25$. The probability to get heads on the third toss is again 0.5; thus, the probability of three heads in a row (no pun intended!) is $0.5 \cdot 0.5 \cdot 0.5 = 0.5^{\text{number of tosses}} = 0.125$.

Let us now assume you lose two out of three tosses. Of course, you would have assumed to win, on average, 1.5 times, half of the three tosses. If your significance level is 0.05, can you accuse me of cheating?

THINK

BREAK

No! As you can see in the fourth column, there are three out of eight results in which heads occurs twice and even four results in which heads occurs two times or more. Thus, the summed probability of these cases is $0.125 + 0.125 + 0.125 + 0.125 = 0.5$, ten times as high as your significance level allows. In fact, you could not even accuse me of cheating if you had lost all three tosses, because even the outcome "three heads" already has a probability higher than 0.05, namely 0.125. This latter situation can be represented graphically in a bar plot as in Figure 5.4, where the x-axis lists the number of heads and the height of the bar indicates the probability of that outcome.

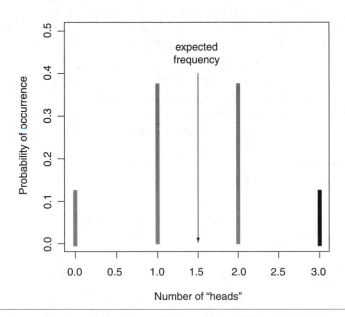

Figure 5.4 All possible outcomes of tossing a coin three times and their summed probabilities (when H_0 is true, one-tailed).

The final important aspect that needs to mentioned now involves the kind of alternative hypothesis. So far, we have always been concerned with directional alternative hypotheses: your alternative hypothesis was "He does not play fair: the probability for heads is *larger* [and not just *different*] than that for tails." The kind of significance test we discussed is correspondingly called a one-tailed test, because we are only interested in one direction in which the observed result deviates from the expected result. Again visually speaking, when you summed up the bar lengths in Figure 5.4, you moved from the null hypothesis expectation in only one direction. This is important, because the decision for or against the alternative hypothesis is based on the cumulative lengths of the bars of the observed result and the more extreme ones in that direction.

However, often we only have a non-directional alternative hypothesis. In such cases, we have to look at both ways in which results may deviate from the expected result. Let us return to the scenario where you and I toss a coin three times, but this time we also have an impartial observer who has no reason to suspect cheating on either part. He therefore formulates the following hypotheses (with a significance level of 0.05):

Statistical H_0: Stefan will win just as often as the other player, namely 50 times.
Statistical H_1: Stefan will win more or less often than the other player.

Imagine now you lost three times. The observer now asks himself whether one of us cheated. As before, he needs to determine which events to consider. First, he has to consider the observed result that you *lost* three times, which arises with a probability of 0.125. But then he also has to considered the probabilities of the other events that deviate from H_0 just as much or even more. With a directional alternative hypothesis, you moved from the null hypothesis only in one direction—but this time there is no directional hypothesis so the observed must also look for deviations just as large or even more extreme in the other direction of the null hypothesis expectation. For that reason—both tails of the distribution in Figure 5.5 must be observed—such tests are called two-tailed tests. As you can see in Table 5.5 or Figure 5.5, there is another deviation from the null hypothesis that is just as extreme; namely that you *win* three times. Since the observer only has a non-directional hypothesis, he has to include the probability of that event, too, arriving at a cumulative probability of 0.125 + 0.125 = 0.25. This logic is graphically represented in Figure 5.5.

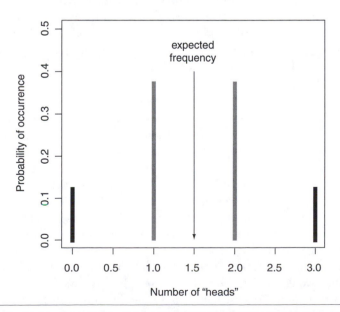

Figure 5.5 All possible outcomes of tossing a coin three times and their summed probabilities (when H_0 is true, two-tailed).

Note that when you tested your directional alternative hypothesis, you looked at the result "you lost three times", but when the impartial observer tested his non-directional alternative hypothesis, he looked at the result "somebody lost three times". This has one very important consequence: when you have prior knowledge about a phenomenon that allows you to formulate a directional, and not just a non-directional, alternative hypothesis, then the result you need for a significant finding can be less extreme than if you only have a non-directional alternative hypothesis. In most cases, it will be like here: the p-value you get for a result with a directional alternative hypothesis is half the p-value you get for a result with a non-directional alternative hypothesis: prior knowledge is rewarded.

With R, this can of course also be done easily for larger numbers of tosses, so let us look at the corresponding result for our first example involving 60 losses in 100 tosses. The significance test now boils down to computing the values for a graph such as Figure 5.6 (using the binomial distribution), and adding up the percentage lengths of the black lines. You would obtain a value 0.02844397 (with `sum(dbinom(60:100,•100,•0.5))`¶), effectively allowing you to adopt your alternative hypothesis.

You should now do "Exercise box 5.3: How often do you . . .?" . . .

Life isn't always this simple, though. We do not always look at scenarios that involve distinct/ categorical states of affairs, and we often look at cases where (multiple) independent variables are involved. In such cases, an exact computation for all possible results is not possible; after all, you cannot compute the likelihood of all averages of subjects and objects because the number of all possible averages is infinite. Thus, many statistical techniques, so-called parametric, asymptotic techniques, are based on approximations. That is, the statistical test does not compute the p-value exactly as we did for two or three heads out of three tosses or 60 heads out of 100 tosses. Rather, it makes use of the fact that, with growing ns, many probability distributions can be approximated very well by functions whose mathematical properties are very well known. For example, with growing n, binomial trials, as the kind of tossing-a-coin experiment we talked about above is usually referred to, can be approximated nearly perfectly by means of a bell-shaped normal distribution, and the corresponding parametric test also says that 60 losses is a significant result. These parametric tests are usually a bit more powerful as well as easier to compute (although with modern-day computing the latter difference is often negligible), but they come at a cost: since they are only approximations that work on the basis of the similarity of your data to a known distribution, they come with a few distributional assumptions, which you must test before you are allowed to apply them. If your data violate these distributional assumptions, you must not compute the parametric/asymptotic test but must resort to a non-parametric, or exact, test of the kind we did for the tosses.

Now that the groundwork has been laid, we will look at a few statistical techniques that allow you to compute p-values to determine whether you are allowed to reject the null hypothesis and, therefore, accept your alternative hypothesis. For each of the tests, we will also look at all relevant assumptions. Unfortunately, considerations of space do not allow me to discuss more than some of the most elementary monofactorial statistical techniques and skip the treatment of many mono-factorial techniques that would be useful or even the multifactorial statistics that a book on statistics in corpus linguistics would undoubtedly deserve (cf. the beginning of this chapter for some references). Since the graphical representation and exploration of distributional data often facilitates the understanding of the statistical results, I will also briefly show how simple graphs are produced. However, again considerations of space do not allow for a detailed exposition so cf. the documentation of the graphics functions I mention, the documentation for par, and the excellent Murrell (2005).

5.2 Categorical Dependent Variables

One of the most frequent corpus-linguistic scenarios involves categorical dependent variables in the sense that one records the frequencies of several mutually exclusive outcomes or categories. We have to distinguish two cases, one in which one just records frequencies of a dependent variable as such without any additional information (such as which independent variable is observed at the same time) and one in which you also include a categorical independent variable; very cursorily, we will also address more complex scenarios involving several independent variables and the case of interval/ratio-scaled independent variables.

5.2.1 One Categorical Dependent Variable, No Independent Variable

Let us look at a syntactic example, namely the word order alternation of English verb–particle constructions as exemplified in (17).

(17)   a. He brought back the book.   Verb Particle Direct Object
       b. He brought the book back.   Verb Direct Object Particle

Among other things, one might be interested in finding out whether two such semantically similar constructions are used equally frequently. To that end, one could decide to look at corpus data for the two constructions. The first step of the analysis consists of formulating the hypotheses to be tested. In this case, this is fairly simple: since the null hypothesis usually states that data are not systematically distributed, the most unsystematic distribution would be a random one, and if sentences are randomly distributed across two construction categories, then both constructions should be equally frequent, just like tossing a coin will in the long run yield nearly identical frequencies of heads and tails. Thus:

H_0:   The frequencies of the two constructions (V DO Part vs. V Part DO) in the population are the same; $n_{V\ DO\ Part} = n_{V\ Part\ DO}$, and variation in the sample is just random noise.

H_1:   The frequencies of the two constructions (V DO Part vs. V Part DO) in the population are not the same; $n_{V\ DO\ Part} \neq n_{V\ Part\ DO}$, and variation in the sample is not just random noise.

Gries (2003a) used a small data sample from the BNC Edition 1 and counted how often each of these two constructions occurred. The simplest descriptive approach is representing these data in a table such as Table 5.6.

Table 5.6 Observed distribution of verb–particle constructions in Gries (2003a)

Verb Particle Direct Object	Verb Direct Object Particle
194	209

We first tell R what the data look like:

```
>•Gries.2003<-c(194,•209)¶
>•names(Gries.2003)<-c("Verb•Particle•Direct•Object",•
  "Verb•Direct•Object•Particle")¶
```

As a first step, we look at the distribution of the data.[4] One simple though not ideal way of representing this distribution graphically would be to use a pie chart. The function to generate a pie chart is pie; the first argument is the vector to be represented as a pie chart:

```
>•pie(Gries.2003)¶
```

Another, probably better possibility is representing the data in a bar plot; the name of the function is barplot.

```
>•barplot(Gries.2003)¶
```

The question that now arises is of course whether this is a result that one may expect by chance or whether the result is unlikely to have arisen by chance and, therefore, probably reflecting a regularity to be uncovered. The actual computation of the test is in this case very simple, but for further discussion below, it is useful to already briefly comment on the underlying logic. The test that is used to test whether an observed frequency distribution deviates from what might be expected on the basis of chance is called the chi-square test. In this case, the null hypothesis states the frequencies are the same so, since we have two categories, the expected frequency of each construction is the overall number of sentences divided by two.

Table 5.7 Expected distribution of verb–particle constructions in Gries (2003a)

Verb Particle Direct Object	Verb Direct Object Particle
201.5	201.5

As with most statistical tests, however, we must first determine whether the test we want to use can in fact be used. The chi-square test should only be applied if:

- all observations are independent of each other (i.e. the value of one data point does not influence that of another); and
- all expected frequencies are larger than or equal to 5.[5]

The latter condition is obviously met, and we assume for now that the data points are independent such that the constructional choice in any one corpus example is independent of other constructional choices in the sample. The actual computation of the chi-square value can be summarized in the equation in (18): you take each observed value, subtract from it the expected value for that cell, square the obtained difference, and divide it by the expected value for that cell again; each of these two elements is sometimes referred to as a contribution to chi-square.

$$(18) \quad \chi^2 = \sum_{i=1}^{n} \frac{(\text{observed} - \text{expected})^2}{\text{expected}} = \frac{(194 - 201.5)^2}{201.5} + \frac{(209 - 201.5)^2}{201.5} \cong 0.558$$

However, since this can become tiresome and prone to errors with larger distributions, and since you would still have to look up this chi-square value to determine whether it is large enough to consider the deviation of the observed values from the expected values significant, we just do the whole test in R, which is extremely simple. Since R already knows what the data look like, we can immediately use chisq.test to compute the chi-square value and the *p*-value at the same time by providing chisq.test with three arguments. The first argument is a vector with the observed data, the second argument is a vector with the probabilities resulting from the null hypothesis; the third argument is correct=T or correct=F: if the size of the data set is small ($15 \leq n \leq 60$), it is often recommended to perform a so-called continuity correction; by calling correct=T you can perform this correction.[6] Since the null hypothesis said the constructions should be equally frequent and since there are just two constructions, this vector contains two times $1 \div 2 = 0.5$:

```
>·chisq.test(Gries.2003,·p=c(0.5,·0.5),·correct=F)¶
·········Chi-squared·test·for·given·probabilities
data:··Gries.2003
X-squared·=·0.5583,·df·=·1,·p-value·=·0.4549
```

The result is clear: the *p*-value is much larger than the usual threshold value of 0.05; we conclude that the frequencies of the two constructions in Gries's sample do not differ from a chance distribution. Another way of summarizing this result would be to say that Gries's data can be assumed to come from a population in which both constructions are equally frequent.

You may now wonder what the *df*-value means. The abbreviation "df" stands for "degrees of freedom" and has to do with the number of values that were entered into the analysis (2 in this case) and the number of statistical parameters estimated from the data (1 in this case); it is customary to provide the *df*-value when you report all summary statistics, as I will exemplify below for each test to be discussed. For reasons of space, I cannot discuss the notion of *df* here in any detail but refer you to full-fledged introductions to statistics instead (Crawley 2002, 2005 are my favorites).

Actually, chisq.test also determines the expected frequencies: chisq.test computes more than it outputs. Since the output of chisq.test is a list, we can first assign that output to a data structure and then inspect the structure of that data structure:

```
>·test<-chisq.test(Gries.2003,·p=c(0.5,·0.5),·correct=F)¶
>·str(test)¶
List·of·8
·$·statistic:·Named·num·0.558
··..-·attr(*,·"names")=·chr·"X-squared"
·$·parameter:·Named·num·1
··..-·attr(*,·"names")=·chr·"df"
·$·p.value··:·num·0.455
·$·method···:·chr·"Chi-squared·test·for·given·probabilities"
·$·data.name:·chr·"Gries.2003"
·$·observed·:·Named·num·[1:2]·194·209
```

```
··..-·attr(*,·"names")=·chr·[1:2]·"Verb·Particle·Direct·Object"·
  "Verb·Direct·Object·Particle"
·$·expected·:·Named·num·[1:2]·202·202
··..-·attr(*,·"names")=·chr·[1:2]·"Verb·Particle·Direct·Object"·
  "Verb·Direct·Object·Particle"
·$·residuals:·Named·num·[1:2]·-0.528··0.528
··..-·attr(*,·"names")=·chr·[1:2]·"Verb·Particle·Direct·Object"·
  "Verb·Direct·Object·Particle"
·-·attr(*,·"class")=·chr·"htest"
```

If we are interested in the expected frequencies, we just need to call that part of the result we are interested in:

```
>·test$expected¶
[1]·201.5·201.5
```

The usual statistical lingo to summarize this result would be something like this: "The verb–particle construction where the particle directly follows the verb occurs 194 times although it was expected 201.5 (i.e. 202) times. On the other hand, the construction with the particle following the direct object was produced 209 times although it was expected 201.5 (i.e. 202) times. This difference, however, is statistically insignificant ($\chi^2 = 0.56$; $df = 1$; $p = 0.455$): we must assume the two constructions are equally frequent in the population for which the sample is representative."

You should now do "Exercise box 5.4: Unidimensional frequency distributions" . . .

For further study/exploration:

- on another way to test whether an observed frequency differs from an expected percentage: ?prop.test¶
- Stefanowitsch (2005, 2006) for an interesting discussion of the relevance of observed vs. expected frequencies

---

5.2.2 One Categorical Dependent Variable, One Categorical Independent Variable

The probably more frequent research scenario with dependent categorical variables, however, is that one records not just the frequency of the dependent variable but also that of an independent categorical variable at the same time. Fortunately, the required method is very similar to that of the previous section: it is called the chi-square test for independence and has the same two requirements as the chi-square test in Section 5.2.1. In order to explore how this test works when an additional variable is added, let us revisit the example of Gries (2003a) from above. The above discussion actually provided only a part of the data. In fact, the above characterization left out one crucial factor, namely one of many independent variables whose influence on the choice of construction Gries wanted to investigate. This independent variable to be looked at here was whether the referent of the direct object was abstract (such as *peace*) or concrete (such as *the book*). Again, we first formulate the hypotheses:

H_0: The frequencies of the two constructions (V DO Part vs. V Part DO, the dependent variable) do not vary depending on whether the referent of the direct object is abstract or concrete (the independent variable).

H_1: The frequencies of the two constructions (V DO Part vs. V Part DO, the dependent variable) vary depending on whether the referent of the direct object is abstract or concrete (the independent variable).

(These are obviously just the text hypotheses, we will turn to the statistical hypotheses presently.) The data Gries (2003a) analyzed resulted in this data set:

Table 5.8 Observed distribution of verb–particle constructions in Gries (2003a)

	V Particle Direct Object	V Direct Object Particle	Totals
Referent of DO = abstract	125	64	189
Referent of DO = concrete	69	145	214
Totals	194	209	403

Before we introduce how the statistical hypotheses are formulated and such data are then tested for significance, let us first represent the data graphically. We first read the data from <C:/_qclwr /_inputfiles/stat_vpc.txt> into R:

```
> Gries.2003<-read.table(choose.files(), header=T, sep="\t", comment.char="")¶
> attach(Gries.2003); str(Gries.2003)¶
'data.frame': 403 obs. of 2 variables:
 $ CONSTRUCTION: Factor w/ 2 levels "V_DO_PART","V_PART_DO": 2 1 1 1 2 1. . .
 $ CONCRETENESS: Factor w/ 2 levels "abstract","concrete": 1 1 2 2 2 2 1. . .
```

Then, now that we know the column names, etc., we use the generic plot command to get a so-called mosaic plot of the data; the first-named variable is used for the *x*-axis.

```
> plot(CONSTRUCTION, CONCRETENESS, xlab="CONSTRUCTION", ylab="CONCRETENESS")¶
```

In this kind of plot, relations of variable levels of one variable are reflected in the column widths while the relations of variable levels of the other variable are reflected using colors. It immediately emerges that the constructional choices are markedly different in the two referent groups. In order to test whether this difference is large enough to reach significance in the present data set, we need to formulate the statistical hypothesis. To that end, it is useful to first consider what the expected frequencies are. However, the computation of the statistical hypotheses is superficially more complex. Since this is such a central question, we will deal with it more thoroughly although R will facilitate things considerably. Let us assume Gries had found the following frequencies of

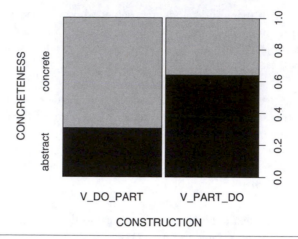

Figure 5.6 Mosaic plot of the distribution of verb–particle constructions in Gries (2003a).

constructions and referents. The letters a, b, c, and d represent any numbers that add up to the specified row and column totals. They are also often used in general to refer to the cells of a 2×2 table such as that instantiated by Table 5.9.

Table 5.9 Fictitious observed distribution of verb–particle constructions

	V Particle Direct Object	V Direct Object Particle	Totals
Referent of DO = abstract	a	b	100
Referent of DO = concrete	c	d	100
Totals	100	100	200

In this case, the computation of the expected frequencies is simple. The null hypothesis usually postulates a random distribution, such that no level is over- or underrepresented. Thus, in this hypothetical case, this would be the distribution following from the null hypothesis.

Table 5.10 Fictitious expected distribution of verb–particle constructions

	V Particle Direct Object	V Direct Object Particle	Totals
Referent of DO = abstract	50	50	100
Referent of DO = concrete	50	50	100
Totals	100	100	200

The marginal totals are all 100, each variable has two equally frequent levels so we just insert 50 everywhere. Thus, the statistical hypotheses appear to be:

H_0: $n_{\text{V DO Part \& Ref DO = abstract}} = n_{\text{V DO Part \& Ref DO = concrete}} = n_{\text{V Part DO \& Ref DO = abstract}}$
$= n_{\text{V Part DO \& Ref DO = concrete}}$

H_1: just like H_0 with at least one unequal sign in it.

However, reality is often more complex because one does not always have data with all identical marginal totals. For example, naturally-occurring data hardly ever result in such homogeneous tables, and even in experiments intended to result in such tables, subjects may fail to provide responses, etc. We therefore now look at Gries's (2003a) real data since here it is not possible to determine the expected frequencies as intuitively as above. For example, it would not make sense to formulate the above statistical null hypothesis because the marginal totals from Gries (2003a) are *not* identical. But if the concreteness of the referent really had no effect, then the frequencies of the two constructions with abstract and concrete referents should always be proportional to the overall frequencies of the constructions in the whole data set. That is, (i) all marginal totals must remain identical—after all, they represent the data that were actually obtained; and (ii) the marginal totals determine the frequencies within the cells of each row and column. From this, a very complex set of statistical null hypotheses follows, which we will simplify in a moment:

$$H_0: \quad n_{\text{V DO Part \& Ref DO = abstract}} : n_{\text{V DO Part \& Ref DO = concrete}} \quad \propto$$
$$n_{\text{V Part DO \& Ref DO = abstract}} : n_{\text{V Part DO \& Ref DO = concrete}} \quad \propto$$
$$n_{\text{Ref DO = abstract}} : n_{\text{Ref DO = concrete}}$$

and

$$n_{\text{V DO Part \& Ref DO = abstract}} : n_{\text{V Part DO \& Ref DO = abstract}} \quad \propto$$
$$n_{\text{V DO Part \& Ref DO = concrete}} : n_{\text{V Part DO \& Ref DO = concrete}} \quad \propto$$
$$n_{\text{V DO Part}} : n_{\text{V Part DO}}$$

$H_1:$ just like H_0 with at least one $!\propto$ in it.

In other words, you cannot just assume that you have $2 \times 2 \, (= 4)$ cells and that, therefore, each expected frequency is approximately 101, namely $403 \div 4$. If you did that, the row totals of the first row would turn out to be approximately 200—but this cannot be true because Gries observed only 189 cases with known referents. Thus, the information that only 189 such cases were observed has to be entered into the computation of the expected frequencies. The easiest way to do this is via percentages. We have $^{189}/_{403}$ cases with abstract referents; the probability of this event is thus 0.469. Also, we have $^{209}/_{403}$ cases of Verb–DirectObject–Particle; the probability of this event is therefore 0.519. If both events are truly random, and thus independent of each other, then the probability of their co-occurrence is $0.469 * 0.519 \approx 0.2434$. Since there are 403 cases with this probability of co-occurrence, the expected frequency is $403 * 0.2434 \approx 98.09$. This seemingly complex logic can be simplified to the formula in (19):

$$(19) \quad n_{\text{expected freq. of a cell}} = \frac{\text{row totals of the cell} * \text{column totals of the cell}}{N} = \frac{189 * 209}{403} = 98.02$$

If you do that for all cells of Gries's (2003a) data, these are the expected frequencies you get (with maybe slight deviations due to rounding).

Table 5.11 Observed distribution of verb–particle constructions in Gries (2003)

	V Particle Direct Object	V Direct Object Particle	Totals
Referent of DO = abstract	90.98	98.02	189
Referent of DO = concrete	103.02	110.98	214
Totals	194	209	403

Note that the ratios of values in each row and each column now correspond to the corresponding ratios in the marginal totals: $90.98:98.02 \propto 103.02:110.98 \propto 194:209$ (for the rows) and $90.98:103.02 \propto 98.02:110.98 \propto 189:214$ (for the columns). We again assume all observations are independent from each other, and since all expected frequencies are larger than 5, all conditions for the chi-square test are met and we can now compute a chi-square test for this data set using equation (18) as before.[7] Also, since the more the observed frequencies deviate from the expected frequencies, the larger the chi-square, we can now always reformulate the statistical hypotheses for frequency data in a very elegant way:

H_0: $\chi^2 = 0$.
H_1: $\chi^2 > 0$.

While this is the procedure underlying chi-square tests of frequency tables, R makes it much easier. The function is again `chisq.test` and the most important arguments it takes are:

- a two-dimensional table for which you want to compute a chi-square test;
- `correct=T` or `correct=F`; cf. above.

We have already loaded the data so we can immediately execute the function:

```
>•(test<-chisq.test(table(CONSTRUCTION,•CONCRETENESS),•correct=F))¶
••••••••Pearson's•Chi-squared•test
data:••table(CONSTRUCTION,•CONCRETENESS)•
X-squared•=•46.1842,•df•=•1,•p-value•=•1.076e-11
```

For the sake of completeness, we also compute the expected frequencies by using the result of the chi-square test again and calling up a part of the output that is normally not provided (by adding $expected to the function call):

```
>•test$expected¶
•••••••••••CONCRETENESS
CONSTRUCTION•abstract•concrete
•••V_DO_Part•98.01737•110.9826
•••V_Part_DO•90.98263•103.0174
```

The results show that the variable CONCRETENESS is highly significantly correlated with the choice of construction. However, we still do not know (i) how strong the effect is and (ii) how CONCRETE-NESS influences the choice of construction. As to the former, it is important to note that one cannot use the chi-square value as a measure of effect size, i.e. as an indication of how strong the correlation between the two investigated variables is. This is due to the fact that the chi-square value is dependent on the effect size, but also on the sample size. We can test this in R very easily:

```
>•chisq.test(table(CONSTRUCTION,•CONCRETENESS)*10,•correct=F)¶
••••••••Pearson's•Chi-squared•test
```

```
data:··table(CONSTRUCTION,·CONCRETENESS)·*·10
X-squared·=·461.8416,·df·=·1,·p-value·<·2.2e-16
```

As is obvious, when the sample is increased by one order of magnitude, so is the chi-square value. This is of course a disadvantage: while the sample size of the table is larger, the relations of the values in the table have of course not changed. In order to obtain a measure of effect size that is not influenced by the sample size, one can transform the chi-square value into a measure of correlation. This is the formula for the measure ϕ (read: phi, for $k \times 2/m \times 2$ tables, where k and m are the numbers of rows and columns respectively) or Cramer's V (for $k \times m$-tables with $k, m > 2$).

$$(20) \quad \phi \text{ or Cramer's } V = \sqrt{\frac{\chi^2}{n * (\min[k, m] - 1)}}$$

The theoretically extreme values of this correlation coefficient are 0 (no correlation) and 1 (perfect correlation).[8] With R, this computation can be done in one easy step, but for expository reasons we break it down into smaller steps:

```
>·numerator<-chisq.test(table(CONSTRUCTION,·CONCRETENESS),·
  correct=F)$statistic¶
>·denominator<-sum(table(CONSTRUCTION,·CONCRETENESS))*
  (min(dim(table(CONSTRUCTION,·CONCRETENESS)))-1)¶
>·fraction<-numerator/denominator¶
>·(phi<-sqrt(fraction))¶
X-squared
0.3385275
```

Thus, the correlation is not particularly strong, but it is highly significant. But where does it come from and how can the results be interpreted? The most straightforward way to answer these questions involves inspecting (i) the so-called Pearson residuals and/or (ii) an association plot. The Pearson residuals indicate the degree to which observed and expected frequencies differ: the more they deviate from 0, the more the observed frequencies deviate from the expected ones. (In fact, if you square them, you get the four contributions to chi-square, whose sum in turn corresponds to the overall chi-square value.) They are generated just like the expected frequencies, by executing the function for the chi-square test again and calling up a part of the output normally not provided:

```
>·chisq.test(table(CONSTRUCTION,·CONCRETENESS),·correct=F)$residuals¶
············CONCRETENESS
CONSTRUCTION··abstract··concrete
···V_DO_Part·-3.435969··3.229039
···V_Part_DO··3.566330·-3.351548
```

This table is interpreted as follows: first, positive values indicate cell frequencies which are larger than expected and negative values indicate cell frequencies which are smaller than expected, and you already know that the more the values deviate from 0, the stronger the effect. Thus, the strongest effect in the data is the strong tendency of abstract objects to occur in the construction where the object follows the particle; the second strongest effect is the dispreference of abstract objects to occur in the other construction, etc. (In this case, all residuals are similarly high, but in tables where this is not so, one could distinguish the cells that matter from those that do not.)

Perhaps a more intuitive approach to the same issue involves a so-called association plot. This simple function requires only one argument, namely a two-dimensional table. Thus:

```
> assocplot(table(CONSTRUCTION, CONCRETENESS))¶
```

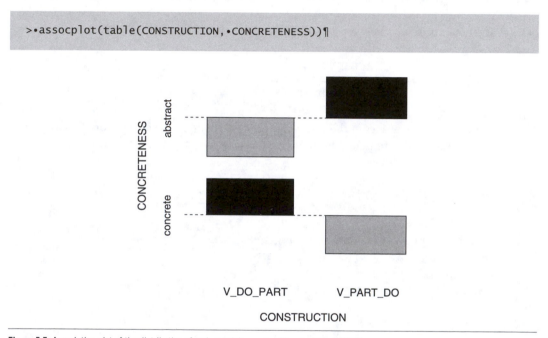

Figure 5.7 Association plot of the distribution of verb–particle constructions in Gries (2003a).

In this representation, black and gray boxes represent table cells whose observed frequencies are greater and smaller than the expected ones respectively (i.e. what corresponds to positive and negative Pearson residuals), and the area of the box is proportional to the difference in observed and expected frequencies.[9]

For further study/exploration: on other ways how to represent contingency tables graphically; cf. ?sieve¶ in the package vcd and ?table.cont¶ in the package ade4

In sum, the data and their evaluations would be summarized as follows: "There is a statistically highly significant albeit moderate correlation between the choice of a verb–particle construction and the abstractness/concreteness ($\chi^2 = 46.18$; $df = 1$; $p < 0.001$; $\phi = 0.34$). The significant result is due to the fact that the construction where the particle follows the verb directly is preferred with abstract objects while the construction where the particle follows the direct object is preferred with concrete objects."

> **You should now do "Exercise box 5.5: Twodimensional frequency distributions" . . .**

Let me briefly mention one important area of application of the chi-square test and other related statistics: measures of collocational strength.[10] Collocational statistics quantify the strength of association or repulsion between a node word and its collocates. Node words tend to attract some words—such that these words occur close to the node word with *greater* than chance probability—while they repel others—such that these words occur close to the node word with *less* than chance probability—while they occur at chance-level frequency with yet others. For example, in Section 2.3 we looked at *alphabetic* and *alphabetical*. Let us return to the collocation of *alphabetical order*. As you can verify with R and the BNC, there are 6,052,200 numbered sentences (occurrences of "<s·n=\"\\d*\" ") and 225 and 16,032 such sentences that contain *alphabetical* tagged as an adjective and *order* tagged as a singular noun, and there are 95 sentences in which both these words occur. This distribution is summarized in Table 5.12.

Table 5.12 Observed distribution of *alphabetical* and *order* in the BNC

	Order	Other words	Totals
Alphabetical	**95**	130	**225**
Other words	15,937	6,036,038	6,051,975
Totals	**16,032**	**6,036,168**	**6,052,200**

Of course, you can evaluate this distribution with a chi-square test; you use the `matrix` function to create a table like Table 5.12 (cf. note 9):

```
>·example<-matrix(c(95,·130,·15937,·6036038),·byrow=T,·ncol=2)¶
>·chisq.test(example,·correct=F)¶
········Pearson's·Chi-squared·test
data:··example
X-squared·=·14993.11,·df·=·1,·p-value·<·2.2e-16
```

As you can see, this distribution is highly significant because the observed co-occurrence of *alphabetical* and *order* (95) is much larger than the expected one (0.6). Two issues have to be mentioned, however. First, the chi-square test is actually not the best test to be applied here given the small expected co-occurrence frequency, which is much smaller than 5. As a matter of fact, there is a huge number of other statistics to quantify the reliability and the strength of the co-occurrence relationship between two (or more) words (cf. Evert 2004, Wiechmann 2008). Second, testing a single collocation like *alphabetical order* is of course not very revealing so, ideally, one would test all collocates of *alphabetical* and all collocates of *alphabetic* to identify semantic or other patterns in the distributional tendencies. We will come back to a similar example below, which is my nice way of saying that you will have to do an assignment involving this approach.

5.2.3 One Categorical Dependent Variable, 2+ Independent Variables

Given considerations of space, we can only discuss methods involving maximally two categorical variables. However, there is a variety of extensions to scenarios where more than two categorical variables are involved. One that is particularly frequent is log-linear analysis, a method which allows to identify those variables and interactions of variables which explain the distribution of the observed frequencies by striking a balance between being most parsimonious and most precise; I personally find some aspects of log-linear analysis relatively challenging for beginners; a textbook containing a rather technical introduction is Agresti (2002).

A method that is less widely used but similar in terms of its analytical power and much simpler is hierarchical configural frequency analysis (H-CFA); cf. von Eye (1990) for a good introduction as well as Gries (to appear) for worked examples with R. H-CFA is basically an extension of chi-square tests to multidimensional tables such that every cell's observed frequency is compared to its expected frequency in a way that is very similar to the inspection of the Pearson residuals or the association plot in the previous section. I recommend this technique especially to those users who find log-linear analysis too daunting.[11]

One of the mathematically simplest approaches involves association rules. Association rules differ from the techniques above in that they, by default at least, do not also provide significance testing. Association rules are conditional probabilities in the form of *if...*, *then...* statements and are used heuristically to determine potentially interesting associations (especially in the domain of data mining: if you ever wondered how Amazon suggests books to you, they use something very similar to association rules; cf. Kolari and Joshi 2004). Important notions in this context are coverage (the percentage of items fulfilling the *if*-antecedent) and support (the percentage of items fulfilling both the *if*-antecedent and the *then*-consequent). Depending on your research objectives, one or more of the above methods may be very useful to you, and I recommend that you try to familiarize yourself with them.

> For further study/exploration: to do log-linear analysis in R, you can use `glm` and/or `loglin`
> NB: Do not use these methods before having consulted the above-mentioned references!

Another useful method that we cannot discuss here is binary or polytomous logistic regression. Logistic regression allows you to predict a categorical dependent variable on the basis of one or more categorical and/or interval/ratio-scaled independent variables.

> For further study/exploration: to do logistic regression in R, you can use
>
> - `glm` for all sorts of generalized linear models
> - `lrm` in the package `Design`
>
> NB: Do not use these methods before having consulted the above-mentioned references!

5.3 Interval/Ratio-scaled Dependent Variables

Another frequent kind of scenario involves interval/ratio-scaled dependent variables. In Section 5.3.1, I will briefly review a few useful descriptive statistics for such dependent variables, but will not discuss significance tests for this scenario (cf. Gries to appear for much discussion and worked examples). In Sections 5.3.2 and 5.3.3, we will then distinguish two cases, one in which you have a two-level categorical independent variable and one in which you have a interval/ratio-scaled independent variable.

5.3.1 Descriptive Statistics for Interval/Ratio-scaled Dependent Variables

First, let us briefly review some useful descriptive statistics with which you can summarize interval/ratio data. Usually, one distinguishes measures of central tendency—what you probably know as averages—and measures of dispersion.

Measures of central tendency serve to summarize the central tendency of an interval/ratio-scaled variable in a single statistic. The most widely known measure of central tendency is the mean. You compute the mean by summing all observations/cases and dividing the sum by the number of observations. In R you use mean, which takes as its only argument a vector of values:

```
>•mean(c(1,•2,•3,•4,•5))¶
[1]•3
```

Despite its apparent simplicity, there are two important things to bear in mind. First, the mean is extremely sensitive to outliers: a single very high or very low value can influence the mean so strongly that it stops being a useful statistic. In the example below, the mean value of 101.7 neither represents the first nine nor the tenth value particularly well.

```
>•mean(c(1,•2,•3,•2,•1,•2,•3,•2,•1,•1000))¶
[1]•101.7
```

In such cases, you should better either trim the data of extremes (using the argument trim) or choose the median as a less sensitive measure of central tendency. If you sort the values in ascending order, the median is the middle value (or the mean of the two middle values). As you can see below, the median of this vector is two, which represents the distribution of the first nine values very well.

```
>•sort(c(1,•2,•3,•2,•1,•2,•3,•2,•1,•1000))¶
•[1]••••1••••1••••1••••2••••2••••2••••2••••3••••3•1000
>•median(c(1,•2,•3,•2,•1,•2,•3,•2,•1,•1000))¶
[1]•2
```

Thus, whenever you look at a distribution as extreme as the above example or whenever your data are better conceived of as ordinal, you should use the median, not the mean.

The second important thing to bear in mind is never, ever(!) to report a measure of central tendency without a measure of dispersion because without a measure of dispersion, one can never know how well the measure of central tendency can in fact summarize all of the data, just as in the above example the mean did not characterize the ten values' central tendency very well.[12] Let us look at an example by comparing two cities' average temperature in one year:

```
>•City1<-c(-5,•-12,•5,•12,•15,•18,•22,•23,•20,•16,•8,•1)¶
>•City2<-c(6,•7,•8,•9,•10,•12,•16,•15,•11,•9,•8,•7)¶
>•mean(City1)¶
[1]•10.25
>•mean(City2)¶
[1]•9.833333
```

From the means alone, it seems as if the two cities have very similar climates, the difference between the means is very small. However, if you look at the data in more detail, especially when you do so graphically, you can see immediately that the cities have very different climates—I know where I prefer to be in February—but just happen to have similar averages:

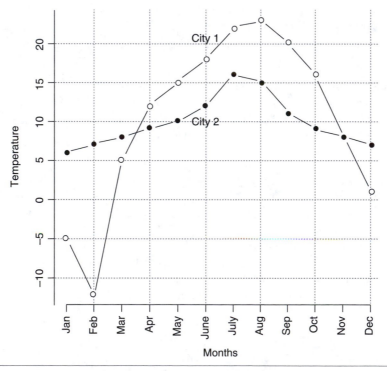

Figure 5.8 Average fictitious temperatures of two cities.

Obviously, the mean of City2 summarizes the central tendency in City2 much better than the mean of City1 because City1 exhibits a much greater degree of dispersion of the data throughout the year. Measures of dispersion quantify this and summarize it into one statistic. The most widely used measure for ratio-scaled data is the so-called standard deviation, which is computed as represented in (21):

$$(21) \qquad sd = \sqrt{\dfrac{\sum\limits_{i=1}^{n} (x_i - \bar{x})^2}{n - 1}}$$

While this may look daunting at first, the "translation" into R (as applied to City1) should clarify this:

```
>•denominator<-length(City1)-1¶
>•numerator<-sum((City1-mean(City1))^2)¶
>•sqrt(numerator/denominator)¶
[1]•11.12021
```

But of course R has a function for this:

```
>•sd(City1)¶
[1]•11.12021
>•sd(City2)¶
[1]•3.157483
```

Note in passing that the standard deviation is the square root of another well-known measure of central tendency, the so-called variance (the R function is var). As is obvious, the measure of dispersion confirms what we could already guess from the graph: the values of City2 are much more homogeneous than the values of City1. It is important, however, to note here that you can only make this kind of statement on the basis of standard deviations or variances when the means of the two vectors to be compared are very similar. This is because the standard deviation is dependent on the size of the mean:

```
>•sd(City1)¶
[1]•11.12021
>•sd(City1*10)¶
[1]•111.2021
```

Thus, when the means of two distributions are markedly dissimilar, then you must not compare their dispersions by means of the standard deviation—rather, you must compare the dispersions on the basis of the variation coefficient, which you get by dividing the standard deviation by the mean:

```
>•sd(City1)/mean(City1)¶
[1]•1.084899
>•sd(City1*10)/mean(City1*10)¶
[1]•1.084899
>•sd(City2)/mean(City2)¶
[1]•0.3210999
```

As you can see, now the measures of dispersion for City1 and its derivative are the same, and they are still considerably higher than that of City2, which is to be expected given the similarities of the means.

Another way of summarizing the dispersion of an interval/ratio variable is again using a measure that is normally used for ordinal variables but helps when you have outliers, etc. (just like the median may replace the mean as a measure of central tendencies in such cases). This measure of dispersion for ordinal variables is the so-called interquartile range (IQR), which gives you the length of the interval around the median that includes about half of the data:

```
>•IQR(City1)¶
[1]•14.5
>•IQR(City2)¶
[1]•3.5
```

Again, the conclusions are the same: City1 is more heterogeneous than City2:

> For further study/exploration:
>
> - on how to get the range of a vector (i.e. a shortcut for max(City1)-min(City1)):
>   ?range¶
> - on how to get the median absolute deviation, another robust measure of dispersion:
>   ?mad¶

A very revealing way to look at such data is the so-called boxplot, a graph that provides a lot of information about the distribution of a vector; cf. Figure 5.9.

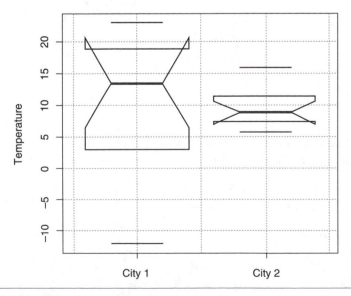

Figure 5.9 Boxplots with notches of two cities.

This plot is generated as follows:

```
>•boxplot(City1,•City2,•notch=T,•ylab="Temperature",•names=c("City•1",•
  "City•2"));•grid()¶
```

This plot tells you something about the central tendency because the bold horizontal lines represent the medians of each distribution. Secondly, the horizontal lines delimiting the boxes at the top and at the bottom extend from the upper to the lower hinge (roughly, the two data points delimiting the highest 25 percent and the lowest 25 percent of all the data). Thirdly, the whiskers—the dashed vertical lines with the horizontal limits—extend to the most extreme data point which is no more than 1.5 times the interquartile range from the box (the default value of 1.5 can be changed; enter ?boxplot¶ at the R prompt). Fourthly, every outlier beyond the whiskers would be represented by one small circle. Finally, the notches extend to ±1.58*IQR/sqrt(n) (enter ?boxplot.stats¶ at the R prompt for information on hinges and whiskers). If the notches of two boxplots do not overlap, this is strong *prima facie* evidence that the medians are significantly different from each other (but of course you would still have to test this properly). The function grid() draws the dashed gray line grid into the coordinate system.

If we apply this to our example, we can now see all of what we have observed so far separately at one glance: the central tendencies of both cities are very similar (because the medians are close to each other and the notches overlap). The first city exhibits much more heterogeneous values than the second city (because the boxes and the whiskers of the first city cover a larger range of the *y*-axis, also, the notches of City1 are huge). Note finally the difference between the top and the bottom of the box in the left plot. While the bottom poses no problems, the top is sometimes difficult to understand because the horizontal line, so to speak, folds back-/downwards. This is R's way of giving you both the correct top end of the box and the full extension of the notch. In this case, the upper end of the box does not go up as far the notch, which is why the line for the notch extends higher than the upper limit of the box.

For further study/exploration:

- on how to compute one-sample *t*-tests and one-sample Wilcoxon signed-rank tests to test whether the central tendency of one dependent interval variable differs from an expected value (cf. below for the corresponding two-sample test): ?t.test¶ and ?wilcox.test¶
- on how to test whether one dependent interval variable is distributed differently from what is expected (cf. below for the corresponding two-sample test): ?ks.test¶

5.3.2 One Interval/Ratio-scaled Dependent Variable, One Categorical Independent Variable

Another very frequent scenario involves the situation in which you have one interval/ratio-scaled dependent variable and one independent categorical variable; the more detailed discussion here will be restricted to the case where the independent categorical variable has just two levels. As an example for this scenario, we can return to the example we discussed above: the different lengths of subjects and (direct) objects. Let us first formulate the hypotheses, assuming we operationalize the lengths by the numbers of syllables:

H$_0$: On average, English subjects are as long as English direct objects; mean length of English subjects=mean length of English direct objects.

H$_1$: On average, English subjects are shorter than English direct objects; mean length of English subjects<mean length of English direct objects.

(We may hypothesize that because referents of subjects are often assumed to be given or accessible information, and given/accessible information is usually encoded with less linguistic material than the new information that objects often provide.) Let us further assume we have investigated so many randomly drawn subjects and direct objects from a corpus until we had 152 instances of each and stored them in a data frame that conforms to the specifications in 5.1.4. We first load the data from <C:/_qclwr/_inputfiles/stat_subjectobject.txt>:

```
>•subj.obj<-read.table(choose.files(),•header=T,•sep="\t",•comment.char="")¶
>•attach(subj.obj);•str(subj.obj)¶
'data.frame':•••304•obs.•of••3•variables:
•$•CASE••••:•int••1•2•3•4•5•6•7•8•9•10•...
•$•LENGTH••:•int••5•1•7•1•1•4•10•1•3•1•...
•$•RELATION:•Factor•w/•2•levels•"object","subject":•2•2•2•2•2•2•2•2•2•2•...
```

Next, we look at the data graphically by again using boxplot. Although this was not mentioned above, there are two different ways of using boxplot. One is to provide the function with vectors as arguments, as we did it above for the two cities. We could do the same here, but would have to make sure we use the right values. In order to stretch the *y*-axis, we can even scale it logarithmically on the *y*-axis (log="y"):

```
>•boxplot(LENGTH[RELATION=="subject"],•LENGTH[RELATION=="object"],•
  notch=T,•log="y")¶
```

However, if the data frame looks as I told you it should, then R also offers the possibility of using a formula notation in which the dependent variable is followed by a tilde ("~") and the independent variable:

```
>•boxplot(LENGTH~RELATION,•notch=T,•log="y");•grid()¶
```

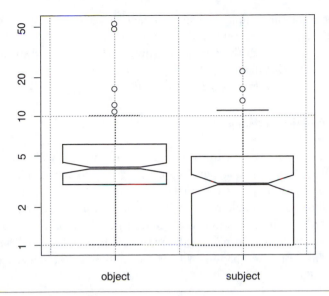

Figure 5.10 Boxplots with notches of two grammatical relations.

As we can see, the median of the objects is slightly larger than that of the subjects, but we can also immediately see that there is a lot of variability. Let us now also compute the means:

```
>•mean(LENGTH[RELATION=="object"])¶
[1]•5.118421
>•mean(LENGTH[RELATION=="subject"])¶
[1]•3.769737
```

or, because you remember `tapply` from Section 3.6.3:

```
>•tapply(LENGTH,•RELATION,•mean)¶
••object••subject
5.118421•3.769737
```

Also, since I told you never to provide a measure of central tendency without a measure of dispersion, we also compute the standard deviations for both:

```
>•tapply(LENGTH,•RELATION,•sd)¶
••object••subject
6.021962•3.381818
```

The usual test for such a case—one dependent interval/ratio variable and one independent nominal variable—is the so-called *t*-test for independent samples. The next step again consists of determining whether we are in fact allowed to perform a *t*-test here. The *t*-test for independent samples is a parametric test and may only be used if:

- the observations of the samples are independent of each other such that there is no meaningful relation between, say, pairs of data points;
- the populations from which the samples are drawn are normally distributed (especially with sample sizes smaller than 30);
- the variances of the two samples are homogeneous.

As to the former assumption, the subjects and objects were randomly drawn from a corpus so there is no relation between the data points. As to the latter two assumptions, boxplots are such useful graphs that we can already guess from them that neither assumption will be met: if the data were perfectly normally distributed, the black line representing the median should be in the middle of the boxes, and the whiskers should extend equally to the top and the bottom. And, if the variances were homogeneous, both boxes should be about equally large. But of course we still need to do the real tests. Fortunately, both these conditions can be tested easily in R. They both come as significance tests in their own right. The former condition, normality, can be tested, among other things, with the Shapiro–Wilk test. The Shapiro–Wilk test computes a test statistic called W and the hypotheses for this test are:

H_0: The data do not deviate from a normal distribution; $W = 1$.
H_1: The data deviate from a normal distribution; $W \neq 1$.

The test is now really simple. The name of the function is `shapiro.test`, and its only argument is the vector to be tested. Thus:

```
>•shapiro.test(LENGTH[RELATION=="subject"])¶
••••••••Shapiro-Wilk•normality•test
data:••LENGTH[RELATION•==•"subject"]
W•=•0.7868,•p-value•=•1.328e-13
>•shapiro.test(LENGTH[RELATION=="object"])¶
••••••••Shapiro-Wilk•normality•test
data:••LENGTH[RELATION•==•"object"]
W•=•0.4516,•p-value•<•2.2e-16
```

The W values are both much smaller than 1 and both p-values are highly significant, indicating that we must accept H_1: the data strongly deviate from the normal distribution required by the t-test.

The latter condition, homogeneity of variances, can be tested with the so-called F-test, which in turn presupposes the normality we just tested with the Shapiro–Wilk test. If the F-test cannot be applied because normality is violated—like here—then you can use the Fligner–Killeen test or, more flexibly, the Ansari–Bradley test. The variances certainly don't look homogeneous—one is more than three times as large as the other:

```
>•tapply(LENGTH,•RELATION,•var)¶
••object••subject
36.26403•11.43669
>•ansari.test(LENGTH~RELATION)¶
••••••••Ansari-Bradley•test
```

```
data:••LENGTH•by•RELATION•
AB•=•13154,•p-value•=•5.165e-05
alternative•hypothesis:•true•ratio•of•scales•is•not•equal•to•1•
```

It does not look like the variances are comparable enough: the *p*-value is highly significant, indicating that we must reject the assumption that the variances are similar enough. Thus, while the *t*-test would have been our first choice, the nature of the data do not allow us to use it.[13] According to the logic from Section 5.1.5 above, we must now therefore use a non-parametric variant, which is the *U*-test. This test does not have assumptions to be tested, but it does not test the data on an interval/ratio level, but on an ordinal scale. Thus, given what we said above in Section 5.3.1, we first compute the measure of central tendency for ordinal data, the median and the corresponding measure of dispersion:

```
>•tapply(LENGTH,•RELATION,•median)¶
•object•subject
••••••4•••••••3
>•tapply(LENGTH,•RELATION,•IQR)¶
•object•subject
••••••3•••••••4
```

Then, we compute the *U*-test. The function for the *U*-test is called `wilcox.test`. Just like `boxplot` and `var.test`, this test allows us to present the data to be tested in the form of vectors or in the form of a formula, as we will see presently. However, it also takes a few additional arguments, some of which need to be mentioned:

- `paired=T` or `paired=F` (the default): for the *U*-test, you always must set this to `paired=F` or leave it out—otherwise you would tell R that your samples are not independent and compute a different test (cf. a statistics textbook on the differences between independent vs. dependent samples and the *U*-test vs. the Wilcoxon test);
- `correct=T` (the default) or `correct=F`, depending on whether you want to apply a continuity correction or not. It is probably safest if you leave set this to `correct=T` or leave this out;
- `alternative="two-sided"` (the default for non-directional alternative hypotheses/two-tailed tests) or `alternative="greater"` or `alternative="less"` for directional alternative hypotheses/one-tailed tests. It is important here to know what the setting of the alternative hypothesis applies to: if you use the vector input, then your setting of `alternative` applies to the first vector; if you use the formula input, then your setting of `alternative` applies to the alphabetically earlier level of the independent variable.

Given all this, what do the functions for the *U*-test have to look like (with a vector input and with a formula input)?

THINK

BREAK

This is what the input could look like:

```
>•wilcox.test(LENGTH[RELATION=="object"],•LENGTH[RELATION=="subject"],•
  paired=F,•correct=T,•alternative="greater")¶
••••••••Wilcoxon•rank•sum•test•with•continuity•correction
data:••LENGTH[RELATION•==•"object"]•and•LENGTH[RELATION•==•"subject"]
W•=•14453,•p-value•=•6.526e-05
alternative•hypothesis:•true•location•shift•is•greater•than•0
>•wilcox.test(LENGTH~RELATION,•paired=F,•correct=T,•alternative="greater")¶
```

Note that, for the sake of explicitness, I have spelt out all the arguments although I could have omitted `paired=F` and `correct=T` because these are the default settings anyway. More importantly, note that since "object" precedes "subject" in the alphabet, both test versions specify alternative = "greater". Thus, this result would be summarized as follows: "The median length of direct objects was 4 syllables (interquartile range: 3) while the median length of subjects was 3 syllables (interquartile range: 4). Since the data violated both the assumption of normality and the assumption of variance homogeneity, a U-test was computed. The U-test showed that the difference between the two lengths is highly significant ($W = 14,453$, $p_{one\text{-}tailed} < 0.001$): in the population of English for which our sample is representative, direct objects are longer than subjects."

While natural linguistic data are only rarely normally distributed, let me also introduce you to the t-test for independent samples here just in case you ever look at a data sample that does not violate the t-test's assumptions. The name of the function is `t.test`, and it takes similar arguments as the U-test:

- `paired=T` or `paired=F` (the default): for the t-test for independent samples, you always must set this to `paired=F` or leave it out;
- `alternative="two-sided"` (the default) or `alternative="greater"` or `alternative ="less"`, which works in the same way as it does for the U-test.

Thus, to use the t-test for this data test—which, again, in reality you must not do since the t-test's assumptions do not hold!—you would enter this:

```
>•t.test(LENGTH~RELATION,•paired=F,•alternative="greater")¶
••••••••Welch•Two•Sample•t-test
data:••LENGTH•by•RELATION
t•=•2.4075,•df•=•237.627,•p-value•=•0.008412
alternative•hypothesis:•true•difference•in•means•is•greater•than•0
95•percent•confidence•interval:
•0.423636••••••Inf
sample•estimates:
•mean•in•group•object•mean•in•group•subject
•••••••••••••5.118421••••••••••••••3.769737
```

In this case, you get a result very similar to that of the U-test but the difference between the means is only *very* significant rather than *highly* significant. If you had illegitimately performed the t-test, this is how you would summarize the data: "The mean length of direct objects was 5.1 syllables (standard deviation: 6.2) while the mean length of subjects was 3.8 syllables (standard deviation: 3.4). A *t*-test for independent samples showed that the difference between the two lengths is very significant ($t = 2.4075$, $df = 237.627$, $p_{\text{one-tailed}} = 0.0084$): in English, direct objects are longer than subjects." Alternatively, you could use the vector-based input:

```
> t.test(LENGTH[RELATION=="object"], LENGTH[RELATION=="subject"],
  alternative="greater")¶
```

You should now do "Exercise box 5.6: Means" . . .

5.3.3 One Interval/Ratio-scaled Dependent Variable, One Interval/Ratio-scaled Independent Variable

The final statistical method we look at is that involving one interval/ratio-scaled dependent variable and one interval/ratio-scaled independent variable. As an example, let us assume that we are again interested in the lengths of XPs. Let us also assume that we generally believe that the best way of operationalizing the length of an element is by counting its number of syllables. However, we may be facing a data set that is so large that we don't think we have the time to really count the number of syllables, something that requires a lot of manual counting.[14] Since you already know how to use R to count words automatically, however, we are now considering to have R assess the XPs' lengths by counting the words and use that as an approximation for the XPs' lengths in syllables. However, we would first want to establish that this is a valid strategy so we decide to take a small sample of our XPs and count their lengths in syllables and lengths in words and see whether they correlate strongly enough to use the latter as a proxy towards the former.

This kind of question can be addressed using a linear correlational measure. Linear correlation coefficients such as r or τ (cf. below) usually range from -1 to $+1$:

- negative values indicate a negative correlation which can be paraphrased by sentences of the form "the more . . ., the less . . ." or "the less . . ., the more . . .";
- values near 0 indicate a lack of a correlation between the two variables;
- positive values indicate a positive correlation which can be paraphrased by sentences of the form "the more . . ., the more . . ." or "the less . . ., the less . . .".

The absolute size of the correlation coefficient, on the other hand, indicates the strength of the correlation. Our hypotheses are therefore as follows:

H_0: The lengths in syllables do not correlate with the lengths in words; $r/\tau = 0$.
H_1: The lengths in syllables correlate positively with the lengths in words such that the more words the XP has, the more syllables it has; $r/\tau > 0$.

Let us now load the data from the file <C:/_qclwr/_inputfiles/stat_lengths.txt> into R:

```
>•lengths<-read.table(choose.files(),•header=T,•sep="\t",•comment.char="")¶
>•attach(lengths);•str(lengths)¶
'data.frame':•••302•obs.•of••2•variables:
•$•LENGTH_SYLL:•int••5•1•7•1•1•4•10•1•3•1•...
•$•LENGTH_WRD•:•int••3•1•3•1•1•3•6•1•2•1•...
```

As usual, we begin by inspecting the data visually. Conveniently, we can again use the generic plot function: just as R recognized factors and generated a mosaic plot in the section on chi-square tests, it recognizes that both vectors contain numeric values and produces the default plot for this case, a scatterplot. As before, the first-named variable is used for the x-axis (the line through the data points summarizes the data points; it has been generated by a separate function, which I will explain at the end of this section).

```
>•plot(LENGTH_SYLL,•LENGTH_WRD,•xlim=c(0,•22),•ylim=c(0,•22))¶
>•grid()¶
```

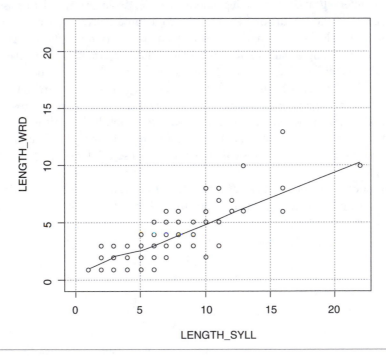

Figure 5.11 Scatterplots of lengths.

It is immediately obvious that there is a positive correlation between the two variables: the larger the syllabic length, the larger the lexical length. The correlation between two interval/ratio-scaled variables is referred to as Pearson's product-moment correlation r (for *regression*), but before we can compute it with R, we again need to determine whether its assumptions are met. Testing r for significance requires that the population from which your samples were taken is bivariately normally distributed, which in practice is often approximated by testing whether each variable is distributed normally. We already know how to do this:

```
>•shapiro.test(LENGTH_WRD)¶
••••••••Shapiro-Wilk•normality•test
data:••LENGTH_WRD
W•=•0.7491,•p-value•<•2.2e-16
>•shapiro.test(LENGTH_SYLL)¶
••••••••Shapiro-Wilk•normality•test
data:••LENGTH_SYLL
W•=•0.8394,•p-value•<•2.2e-16
```

As is easy to see, both variables are not normally distributed—the *p*-values are small and indicate a deviation from normality—so we cannot compute *r*. The solution is again to use a non-parametric alternative to the test we originally wanted to do. Instead of *r*, we now compute another correlation coefficient, Kendall's τ (read: tau). Fortunately for us, the function for both correlations is the same, cor.test. It takes the two vectors to be correlated (the order doesn't matter) and an alternative argument, which is "greater" if your alternative hypothesis postulates a positive correlation (as it does in our example), "less" if your alternative hypothesis postulates a negative correlation, and "two-sided" if your alternative hypothesis does not postulate a particular direction. Thus, recalling the nature of our alternative hypothesis, we write:

```
>•cor.test(LENGTH_SYLL,•LENGTH_WRD,•method="kendall",•alternative="greater")¶
••••••••Kendall's•rank•correlation•tau
data:••LENGTH_SYLL•and•LENGTH_WRD
z•=•16.8139,•p-value•<•2.2e-16
alternative•hypothesis:•true•tau•is•greater•than•0
sample•estimates:
••••••tau
0.7657678
```

It turns out that the correlation is rather high and highly significant so that we might want to be confident that the length in syllables can be reasonably well approximated by the length in words that is computationally easier to obtain. We can thus say: "There is a highly significant positive correlation between the lengths of XPs in words and the lengths of XPs in syllables (Kendall's $\tau = 0.766$, $z = 16.8139$; $p_{\text{one-tailed}} < 0.001$)." Just so that you know in case you have data that are normally distributed: the computation of *r* is just as easy because you only need to change "kendall" into "pearson".

Let me finally mention two things. First, note again that this section only dealt with linear correlations. If you want to do other regressions, you cannot reasonably use cor.test. Secondly, as I mentioned above, the line through the data points summarizes the distribution of the data points. To generate such a (non-parametric robust smoother) line, you just write the following after you have generated the scatterplot:

```
>•lines(lowess(LENGTH_WRD~LENGTH_SYLL))¶
```

(More generally, the variable that is on the *y*-axis in your scatterplot must precede the tilde, the one that is on the *x*-axis must follow the tilde.) I will not explain this any further because the math underlying this curve goes beyond what even most introductions to statistics discuss. This only serves the expository purpose of facilitating the recognition of patterns in the data.

> For further study/exploration: to do simple regressions in R—linear and otherwise—you can use `lm` and/or `rq` from the package `quantreg` or `ols` from the package `Design`
>
> NB: Do not use these methods before having consulted the above-mentioned references!

5.3.4 One Interval/Ratio-scaled Dependent Variable, 2+ Independent Variables

If a research design involves more than one independent categorical variable, then most studies use the probably most frequent statistical technique, analysis of variance, or ANOVA for short. An ANOVA allows you to test which of the 2+ independent variables and which of their interactions have a significant effect on your dependent variable. In the case of one independent variable with two levels, an ANOVA will yield the same result as the *t*-test for independent samples discussed above.

ANOVA is an extremely powerful and versatile method, but this power comes at the cost that especially more complex designs require quite some familiarity with the method, its requirements, and the various kinds of results it can yield. Before you apply it, you should do some serious reading on both the theory behind the method and its implementation.

> For further study/exploration: to do ANOVAs in R, you can use:
>
> - the functions `anova`, `aov` in the base package as well as `Anova` in the package `car`
> - the functions `oneway.test` or `kruskal.test` for non-parametric alternatives to `anova`
> - the functions `lm` or `ols` (from the package `Design`) for all sorts of linear models
> - `plot.design` to graphically represent differences between means
> - the functions `TukeyHSD` and `pairwise.t.test` for *post hoc* tests
>
> NB: Do not use these methods before having consulted the above-mentioned references!

If you have more than one independent interval/ratio-scaled variable, you analyze your data using multiple regression analysis. If you also have one or more independent categorical variables, you use so-called analysis of covariance (or ANCOVA). Just as with ANOVA, these are relatively complicated methods.[15]

> For further study/exploration: to do multiple regression or ANCOVA in R, you can use `lm` for all sorts of linear models. A set of new and very powerful methods, which are fairly complicated, is available with the functions `lmer` and `glmer` (in the package `lme4`; cf. Baayen 2008: Chapter 7)
>
> NB: Do not use these methods before having consulted the above-mentioned references!

While all of the methods so far have been concerned with *testing* hypotheses, there are also cases where one may want to use statistical techniques to *generate* hypotheses or detect patterns in data sets that are so large and/or complex as to rule out a useful manual inspection of the data. Two of the most powerful techniques worth exploring in this context are (hierarchical) cluster analysis and classification and regression trees (CART).

For further study/exploration:

To do cluster analyses in R, you can use:

- the functions `dist` and `hclust`
- the functions `agnes`, `diana`, and `pam` in the package `cluster`

To do CART analysis, you can use `tree` in the package `tree`
To do general data exploration, you can use the package `ggobi`

NB: Do not use these methods before having consulted the above-mentioned references!

5.4 Customizing Statistical Plots

Many of the above sections have involved functions to generate various plots. While there is no space to discuss the huge number of ways in which R allows you to customize these and other graphs, I would like to briefly give an overview of the main arguments you can use to tweak your graphs. For example, when R generates plots, it applies default settings to determine the limits of the axes, the labels of the axes, etc. You can override these by setting them manually:

- `xlab="..."` and `ylab="..."` allow you to specify the labels for the *x*-axis and the *y*-axis;
- `xlim=c(xmin,•xmax)` and `ylim=c(ymin,•ymax)` allow you to specify the range of the *x*-axis and the *y*-axis; for example, `xlim=c(0,•10)` gives you an *x*-axis that begins at 0 and ends at 10;
- `col=c(...)` allows you to specify colors (enter `colors()¶` at the R prompt for a list of color names you can use);
- `text(x,•y,•labels="...")` allows you to position the specified string in the labels argument at the *x* and *y* coordinates given as arguments to `text`; for example, `text (1,•2,•labels="example")` positions (centrally) the word "example" at *x* = 1 and *y* = 2 in the plot.

For more options, cf. the references mentioned at the beginning of this chapter.

5.5 Reporting Results

The previous sections have introduced you to a few elementary statistical methods that are widely applied to corpus data. For each of the methods, I have also shown you how the data would be summarized both in terms of graphic representation and in terms of statistical testing. However, now that you know something about each aspect of the empirical analysis of language using corpus data, I would also like to at least briefly mention the overall structure of an empirical paper that I find most suitable and that corresponds to the structure of quantitative/empirical data in most sciences. This structure looks as represented in Table 5.13.

Table 5.13 Structure of a quantitative corpus-linguistic paper[16]

Part	Content
Introduction	what is the question?
	motivation of the question
	overview of previous work
	formulation of hypotheses
Methods	operationalization of variables
	choice of method (e.g., diachronic vs. synchronic corpus data, tagged vs. untagged corpora, etc.)
	source of data (which corpus/corpora?)
	retrieval algorithm or syntax
	software that was used
	data filtering/annotation (e.g. how were false hits identified, what did you do to guarantee objective coding procedures? how did you annotate your data? etc.)
	choice of statistical test
Results	summary statistics
	graphic representation
	significance test: test statistic, degrees of freedom (where available), and p
	effect size: the difference in means, the correlation, etc.
Discussion	implications of the findings for your hypotheses
	implications of the findings for the research area

This structure is very rigid, but given its wide distribution, you will get used to it very quickly and other people will find your papers well structured.

Also, the structure may seem excessively precise in that it requires you to specify even the retrieval algorithm or syntax and the software that you used. The reason for this is one I already mentioned a few times: different applications incorporate different search parameters: what is a word for MonoConc Pro 2.2 need not be a word for WordSmith Tools 4 or any other concordancer. In fact, in one of the case study assignments, you will see an example where the default settings of different programs produce remarkably different output for the simplest possible kind of concordance, namely a case-insensitive search for one word. Thus, in order for others to be able to understand and replicate your results, you must outline your approach as precisely as possible.

Of course, if you use R, all this is unproblematic because you can just provide your regular expression(s) or even your complete program as pseudocode. Pseudocode is a kind of structured English that allows you to describe what a program/script does without having to pay attention to details of the programming language's syntax (cf. the very useful page by John Dalbey (2003) at <http://www.csc.calpoly.edu/~jdalbey/SWE/pdl_std.html> for some more details). For example, the pseudocode for the script discussed in Section 4.2.2 could look like this:

```
01  clear memory
02  choose corpus files to be searched
03  for each corpus file
04      load the file
05      if the corpus file is from the right register
06          downsize the file to all lines with sentence numbers
```

```
07              retrieve all matches (i.e. lines containing right as an adjective) from
                      the file
08              store all matches with the names of the files successively
09    endif
10 end for
11 insert tab stops for better output
12 delete unwanted tags and unnecessary spaces
13 output result into a file
```

If you provide this information together with the actual perl-compatible regular expressions you used to:

- check the right register line line 5 ("<teiHeader type.*S conv</classCode>");
- downsize the file in line 6 ("<s·n=");
- retrieve the matches in line 7 ("<w·AJ0(-AV0)?>right\\b");
- delete unwanted tags and unnecessary spaces in line 12 ("<.*?>" and "·*\t·*").

Then you will be more explicit than many corpus linguists are at present and than I myself have been on too many occasions in the past.

Now we are at the point where we have covered nearly all the knowledge you can finally get down to business and deal with the more advanced case studies.

6

Case Studies and Pointers to Other Applications

> I think computer searching of corpora is the most useful tool that has been provided to the grammarian since the invention of writing.
>
> (Pullum 2006: 39)

6.1 Introduction to the Case Studies

Now that you have mastered all the basics—the three basic corpus-linguistic methods, different corpora and their annotation formats, as well as how to generate frequency lists, concordances, and collocation displays for various kinds of raw and annotated corpora—it is finally time to look at more realistic corpus-linguistic case studies and, therefore, more comprehensive tasks. On the companion website to the book at <http://groups.google.com/group/corpling-with-r/web /quantitative-corpus-linguistics-with-r>, you will find a variety of case studies that aim at bringing together the various things explained and exemplified in previous chapters. These case studies, which you do not have to do in any particular order, can be distinguished along two dimensions. The first of these is the linguistic sub-discipline, and, as you can see, I will use examples from morphology, syntax, semantics/lexicography, pragmatics/text linguistics, and the dustbin class of "other applications." The second dimension is that of the motivation underlying the case studies. Some of the studies aim at replicating or extending previous studies such that you will do data retrieval, annotation/coding, and some statistical analysis to practice all aspects of quantitative corpus linguistics. Another set of studies is mainly concerned with particular issues of data retrieval and preparation for case studies for which I will provide some theoretical background, but which cannot be simulated here in more detail than you will find below. Finally, there are a few case studies which in fact do not come with a particular theoretical question but which have been included to show you how to do corpus-linguistic things with R for which most people use specific software. Thus, the motivation of these sections is to broaden your horizon beyond what you may think is the limit of what can be done and hopefully stimulate further ideas for exploration; for some case studies, you will also find recommendations for study/exploration.

The case study assignments usually provide a small introduction, sometimes with a few examples and references to earlier work or other software, and then you will find working assignments with explicit instructions as to how to tackle the problem outlined in each section. The idea is that you write one or more scripts that perform some retrieval and analysis operations; I would strongly recommend that you do this in Tinn-R, which will facilitate your programming task and will make it easier to run tests and do debugging. You will find all input files in <C:/_qclwr/_inputfiles>, and as before file names either name the corpora from which the data are taken or in which section a file is supposed to be used.

When you think you have completed all assignments of a section, save the script a final time, run it, and save your output into <C:/_qclwr/_myoutputfiles>. Then, check your solution:

- by comparing your script to my script, which you will find in <C:/_qclwr/_scripts>;
- by comparing your output to the output files resulting from my script, which you will find in <C:/_qclwr/_outputfiles>.

Also, when you run into problems with an assignment such that you cannot put together a regular expression that you think you need, you cannot generate a table the way you think you need it, etc., also turn to the answer key scripts for help. Note, however, that I may have chosen an approach to the problem that differs from yours, so don't just look at the one or two lines where you suspect your problem may be solved—you may have to go over larger parts of my answer key to make sure you understand it alright. If you find a solution to your problem, first try to continue to work on your approach—only switch back to mine again if you can't make any further progress with your strategy.

Finally, note that the goal is not that you guess my approach. The first goal is that you find any one workable solution. Once you have found one solution that works—be it slow and ugly and ...—you can then work on making it faster, more memory-efficient, more elegant, more extensible (in the sense that the script can be easily adapted and used for other data, other searches, etc.); do not forget that there is not just one way of tackling these problems. Do not despair if you don't find a solution or if your solution does not match the one I used, and celebrate if your solution is ultimately better, shorter, more elegant, faster, ... than mine. The important things are that:

- you slowly get your script to do what it is supposed to do;
- you see why your script does what it does;
- you see why the script proposed here does what it does;
- you slowly begin to think like a programmer by compartmentalizing big problems into small ones and seeing creative solutions.

Good luck and have fun with the case studies!

6.2 Some Pointers to Further Applications

While the current version of the companion website is restricted to just a few examples from some areas of linguistics, corpus-linguistic methods of the above kind have much more to offer to the field, and I intend to upload more and more assignments as time permits it. Obviously, *any* field of linguistics in which frequency of (co-)occurrence or the qualitative analysis of naturally occurring data may be useful can benefit strongly from corpus-based methods such as concordances. In what follows, I will mention a few more examples from the above areas and others to briefly illustrate the range of problems eligible to corpus-based methods (without R code or data, though). My overview is highly selective and tainted by work that I found inspiring: I lay no claim to comprehensively surveying the fields I mention—after all, each of these fields merits separate textbooks.

One area within linguistics to which corpus linguistics has a somewhat natural affinity is *text linguistics*. We had a look at two rather specialized applications above, but text linguistics is of course a vast field and I would like to mention briefly two additional applications I find most interesting. One remarkable strand of research is the multidimensional analysis approach by Biber (1988), an approach in which multiple linguistic properties of texts are investigated simultaneously. Rather than simply *assuming* top-down that particular registers or genres are correlated with particular linguistic characteristics, Biber adopted a bottom-up approach: he annotated various texts from different genres, registers, and varieties for many different linguistic properties and used exploratory factor analysis to determine which of the linguistic factors are strongly correlated. As a result, Biber arrives at an inventory of textual properties that—since they are derived rather directly from the data—are much more likely to be more revealing than categories defined a priori.

Another highly interesting approach is TextTiling (cf. Hearst 1997), a technique from the gray area between corpus linguistics and computational linguistics which serves to automatically divide texts into topically coherent multiparagraph units. The algorithm takes as input a text and goes through it to determine how similar two adjacent blocks of texts are in terms of the number of words they share: the larger the number of shared words, the more likely it is that the two blocks of texts are about the same topic. The result is a list of adjacent blocks which are topically related. Hearst shows that her method handles issues of both text summarization and information retrieval successfully, and other scholars have applied the method to stylistic analysis and the alignment of multilingual corpora, a quasi-prerequisite for multilingual corpus analysis and translation studies. A slightly similar approach by Youmans (1990, 1991) has been applied to the study of literature and discourse and involves, among other things, the computation of type–token ratios: in the simplest case, one just goes through a text word by word and stores for each such position the number of types and the number of different tokens. The quotient of these two figures is then shown in a scatterplot for each position in the text and shows how the number of different words in the text develops throughout the text. As is probably obvious from this characterization, you can program this in R with just a few lines.

A somewhat closer connection of corpus linguistics to *discourse and pragmatics* is exemplified by work on the interplay between syntax and pragmatics. For example, much work by Givón and his collaborators on the distribution of given–new information in texts has operationalized the degree to which a referent is new/given and important/unimportant in a particular stretch of discourse by looking at how often this referent is mentioned in a user-defined amount of cotext, which should immediately remind you of concordancing (cf. Givón 1992 for exemplification). Other discourse-functional studies investigate how particular words or constructions are used in authentic contexts (cf., e.g. Tao and Meyer 2006; Tao 2007; Gries and David 2007). Also, corpus-linguistic methods are also gaining ground rapidly in *cognitive linguistics*. In cognitive linguistics, Stefanowitsch and Gries (2006) and Gries and Stefanowitsch (2006) contain a variety of papers on conceptual mappings and the syntax–lexis interface and the number of corpus-based presentations at recent cognitive-linguistic conferences has increased sharply.

Corpus linguistics has also become very influential in *applied linguistics*. Apart from the above issue of measuring the learners' vocabulary, corpus linguists have also taken interest in curriculum development. Some studies have shown that the language represented in teaching materials often deviates significantly from language as found in corpora. Thus, some scholars argue that the development of teaching materials should be based, among other things, on corpus frequencies such that language learners get to learn the most frequent words and their most frequent meanings, the most frequent tenses, the most frequent syntactic structures, etc. first in order to accumulate more useful knowledge as early as possible; cf. Conrad (2000), Fox (1998), Stevens (1991), and for excellent book-length treatments, Hunston (2002), Sinclair (2004), or O'Keeffe, McCarthy, and Carter (2007). For an overview of corpus-based methods in SLA, cf. Gries (2008a).

Corpus-based methods and findings are rapidly gaining ground in the domain of *psycholinguistics*. As mentioned above, psycholinguists often use frequency information derived from corpora, but there are also studies which incorporate much more of the above methods. One out of many examples is recent work on disfluencies in language production. For example, Fox Tree and Clark (1997) use concordancing the London–Lund corpus to identify and analyze disfluencies such as the use of [ðiː] instead of [ðɪ] in NPs; in a later study, Clark and Fox Tree (2002) also study *uh/er* vs. *uhm/erm* (in the London–Lund corpus, the Switchboard corpus, and other corpora). Both studies were concerned with learning about the how speakers plan their speech and monitor the execution of their plans as well as determine whether these expressions actually constitute signals in their own right in much the same way as 'regular words' do.

A second example is concerned with syntactic, or structural, priming or persistence. Syntactic priming refers to the fact that speakers have a tendency to re-use grammatical patterns they have just produced or comprehended shortly ago. While most work in this area has been based on carefully constructed experiments beginning in the 1980s, there is early corpus-based work from that time (e.g. Sankoff and Laberge 1978; Estival 1985) as well as more recent work (Gries 2005a; Szmrecsanyi 2005), which is both more comprehensive and quantitatively more sophisticated. Such approaches are a nice illustration of how the authenticity of corpus data can yield useful results, even in the domain of online utterance processing.

The area of psycholinguistic research that has probably benefited most from corpus-based methods is language acquisition research. I already mentioned the huge CHILDES database of acquisition corpora with transcripts of thousands of hours of recordings and we had a look at the computation of MLU values. One very recent corpus-based study that is strongly based on corpus analysis is Diessel and Tomasello (2005). They investigate children's acquisition of the word order alternation known as particle placement (cf. Section 4.2.3.3, Section 5.2, and (23) for yet another example) studying data on two children from CHILDES.

(23) a. He picked up the book.
      b. He picked the book up.

Inspecting and coding the concordance lines for various factors suspected to govern the alternation, they identify the most relevant factors influencing children's choice of word order and identify a variety of interesting commonalities and differences to findings on adult speech.

Yet another subdiscipline that can benefit a lot from the above methods is the analysis of language development and grammaticalization in *historical linguistics*. To give but one example, determining how lexical words grammaticalize into function words requires retrieving word forms from diachronic corpora and analyzing their uses to determine the frequencies with which words become increasingly more frequent as grammatical markers.

A final interesting area to be mentioned here is the application of corpus-linguistic methods to *authorship attribution/stylometry and the study of literature*. Authorship attribution in the context of forensic linguistics serves to determine who out of several likely authors has actually written a particular text, an issue that may be important in the analysis of ransom letters; a less critical context of application would be the attribution of historical writings to one of several writers. Research has shown that different people tend to have different favorite syntactic structures, words, or kinds of words, which may be so strongly reflected in word frequencies, part-of-speech frequencies, or even letter frequencies that authorship attribution achieves high levels of accuracy; all these preferences can of course be easily determined using the above methods. More complex models can further increase performance. In the domain of literary studies, corpus-based methods can be used to identify turning points or climaxes as well as revealing lexical choices of authors in literary prose; cf. Oakes (1998: Chapter 5) and Juola (2008) for overviews.

Apart from all these disciplines, and many more, *corpus linguistics* itself is also evolving further. For example, there is quite some work on the issue of corpus homogeneity as measured by word frequency lists (cf., e.g. Kilgarriff 2001, 2005) or by more differentiated and grammatically relevant measures (cf. Gries 2005b, 2007a). In addition, the amount of work investigating the convergence of corpus-based and experimental evidence is growing (cf. Keller 2000; Gries 2002a; Tummers, Heylen, and Geeraerts 2005; several papers in the special issue of *Lingua* 115.11 in 2005; Gries, Hampe, and Schönefeld, 2005, to appear; Hoffmann 2006; Arppe and Järvikivi, 2007; to name but a few), and in all these areas R can be extremely helpful or even be necessary (because the operations involved are too numerous or complex to be done by humans alone).

Another highly relevant area of application that has not been mentioned much is that of corpus compilation. This book has mostly dealt with examples where corpus data already existed before R was used. However, some of the assignments have indicated how R can be used to compile and clean data from websites. In addition, it is worth pointing out that you can also use R to annotate a corpus. For example, imagine you have a text file you want to annotate semi-automatically. You could write a script that loads the file to be annotated, identifies each instance of a word you want to annotate, and prompts you to provide an annotation, which then gets saved into the file, too. As a result, the text file will be contain your annotation. For a similar scenario at the MPI-EVA, my script then even remembered previous annotations and suggested them to the annotator to speed up things. Similar examples abound . . .

This was just a very selective compilation; there are many more areas in linguistics integrating, and benefiting from, corpus-based work (e.g. work on subcategorization preferences, distributional acquisition of words, etc.) which I cannot list here individually, but I hope the above examples have given you a taste of what can be done. More generally, it is impossible to even try to anticipate the range of corpus-linguistic applications to which R can be put, but if this book has provided you with some first knowledge of what can be done and, more importantly, also stimulated you to be creative and explore the possibilities, then it has fulfilled its objectives.

Appendix

Websites

The companion website for this book: <http://groups.google.com/group/corpling-with-r/web/quantitative-corpus-linguistics-with-r>

The newsgroup on corpus linguistics with R that hosts the companion website: <http://groups.google.com/group/corpling-with-r>

Corpus Linguistics Overviews

<http://devoted.to/corpora>
<http://www.bmanuel.org/index.html>

Corpus-linguistic References

<http://ifa.amu.edu.pl/~kprzemek/biblios/corpling.zip>

Corpora/Corpus-related Databases

BNC and its versions: <http://www.natcorp.ox.ac.uk/>
BNC Version 1 frequency lists: <http://www.kilgarriff.co.uk/bnc-readme.html>
Brown: <http://icame.uib.no/brown/bcm.html>
CELEX: <http://www.ru.nl/celex/>
CHILDES corpora (and CLAN): <http://childes.psy.cmu.edu>
Christine: <http://www.grsampson.net/RChristine.html>
ICE: <http://www.ucl.ac.uk/english-usage/ice/>
Leipzig Corpora Collection Download Page: <http://corpora.uni-leipzig.de/download.html>
NLTK: <http://nltk.sourceforge.net/>
Susanne: <http://www.grsampson.net/RSue.html>
Switchboard corpus: search the catalog at <http://www.ldc.upenn.edu/> for various versions
The JRC-ACQUIS Multilingual Parallel Corpus, Version 3.0: <http://langtech.jrc.it/IRC-Acquis.html>
Treebanks: <http://www.cis.upenn.edu/~treebank/>

Tags

British National Corpus World Edition: <http://www.comp.lancs.ac.uk/ucrel/claws5tags.html>
Brown corpus: <http://www.comp.leeds.ac.uk/amalgam/tagsets/brown.html>

Journals and Databases

ACL Anthology: <http://acl.ldc.upenn.edu/>
Citeseer: <http://citeseer.ist.psu.edu/>
Computational Linguistics: <http://mitpress.mit.edu/catalog/item/default.asp?ttype=4&tid=10>
Computer Speech and Language: <http://www.elsevier.com/wps/find/journaldescription.cws_home/622808/description#description>
Corpus Linguistics and Linguistic Theory: <http://www.degruyter.com/journals/cllt>
Corpora: <http://www.eup.ed.ac.uk/journals/content.aspx?pageId=1&journalId=12505>
Empirical Language Research: <http://www.ejournals.org.uk/ELR/>
ICAME Journal: <http://nora.hd.uib.no/journal>
International Journal of Corpus Linguistics: <http://www.benjamins.com/cgi-bin/t_seriesview.cgi?series=IJCL>
Literary and Linguistic Computing: <http://llc.oxfordjournals.org/>
Language Resources and Evaluation (formerly known as Computers and the Humanities): <http://www.springerlink.com/link.asp?id=113189>
Web-as-corpus 1: <http://www.webcorp.org.uk/>
Web-as-corpus 2: <http://wacky.sslmit.unibo.it/doku.php>

Text Collections and Other Relevant Projects

Leipzig corpora download page: <http://corpora.informatik.uni-leipzig.de/download.html>
Marina Santini: <http://www.nltg.brighton.ac.uk/home/Marina.Santini/>
Project Gutenberg: <http://www.gutenberg.org/>
University of Virginia Etext Center: <http://etext.lib.virginia.edu/collections/languages/>

R and R Add-ons (Open Source)

R Project: <http://www.r-project.org>
R Project download page CRAN: <http://cran.r-project.org/mirrors.html>
R websites: <http://www.rseek.org/>, <http://wiki.r-project.org/>, and <http://google.com/coop/cse?cx=018133866098353049407%3Aozv9awtetwy>
Tinn-R: <http://www.sciviews.org/Tinn-R/>
SciViews package: <http://www.sciviews.org/SciViews-R/>
R Graphics: <http://addictedtor.free.fr/graphiques/>
R sitesearch toolbar for Firefox: <http://addictedtor.free.fr/rsitesearch/>
Editors for R: <http://www.sciviews.org/_rgui/projects/Editors.html>
R reference card (short): <http://cran.r-project.org/doc/contrib/Short-refcard.pdf>
R reference card (long): <http://cran.r-project.org/doc/contrib/refcard.pdf>
R/Rpad reference card: <http://www.rpad.org/Rpad/Rpad-refcard.pdf>

Mailing lists on using R for general statistical evaluation (i.e. not necessarily corpus-linguistic data) in linguistics:

R-lang: <https://ling.ucsd.edu/mailman/listinfo.cgi/r-lang>
StatForLing with R: <http://groups.google.com/group/statforling-with-r> (this is the newsgroup hosting the companion website for Gries's textbook on statistics for linguistics with R)

Office Software, Text Editors, and Other Useful Software (Mostly Open Source)

Openoffice.org: <http://www.openoffice.org>
Unicode consortium: <http://unicode.org/>
HTML code strippers: <http://personal.cityu.edu.hk/~davidlee/devotedtocorpora/software.htm#HTMLcodestrippers>
Web snaggers: <http://personal.cityu.edu.hk/~davidlee/devotedtocorpora/software.htm#Web Snaggers>

Regular Expressions

http://regexlib.com/
http://regexadvice.com/blogs/ssmith/contact.aspx
Regulazy: <http://tools.osherove.com/CoolTools/Regulazy/tabid/182/Default.aspx>
The Regulator: <http://tools.osherove.com/CoolTools/TheRegulator/tabid/185/Default.aspx>

References

Agresti, Alan. 2002. *Categorical Data Analysis*. Hoboken, NJ: Wiley.

Aijmer, Karin. 1984. *Sort of* and *kind of* in English conversation. *Studia Linguistica* 38/2: 118–128.

——. 1986. Discourse variation and hedging. In *Corpus Linguistics II: New Studies in the Analysis and Exploitation of Computer Corpora*, ed. Jan Aarts and Willem Meijs, 1–18. Amsterdam: Rodopi.

Arppe, Antti and Juhani Järvikivi. 2007. Every method counts: combining corpus-based and experimental evidence in the study of synonymy. *Corpus Linguistics and Linguistic Theory* 3/2: 131–159.

Aston, Guy (ed.). 2001. *Learning With Corpora*. Bologna. CLUEB.

Baayen, R. Harald. 2008. *Analyzing Linguistic Data: A Practical Introduction to Statistics Using R*. Cambridge: Cambridge University Press.

Baayen, R. Harald, Richard Piepenbrock, and Leon Gulikers. 1995. *The CELEX Lexical Database* (Release 2). Philadelphia, PA: Linguistic Data Consortium.

Baker, Paul, Andrew Hardie, and Tony McEnery. 2006. *A Glossary of Corpus Linguistics*. Edinburgh: Edinburgh University Press.

Barnbrook, Geoff. 1996. *Language and Computers*. Edinburgh: Edinburgh University Press.

Baron, Faye and Graeme Hirst. 2003. Collocations as cues to semantic orientation. Unpublished MS. Downloaded from <http://citeseer.ist.psu.edu/rd/0%2C683844%2C1%2C0.25%2CDownload/http%3AqSqqSqftp.cs.toronto.eduqSqpubqSqghqSqBaronqPqHirst-2003.pdf> on September 20, 2006.

Baroni, Marco and Silvia Bernardini (eds.). 2006. *Wacky! Working papers on the web as corpus*. Bologna: GEDIT. [ISBN 88-6027-004-9] Downloadable from <http://wackybook.sslmit.unibo.it/>

Beal, Joan C., Karen P. Corrigan, and Hermann L. Moisl (eds.). 2007a. *Creating and Digitizing Language Corpora: Synchronic Databases Vol. 1*. Basingstoke: Palgrave Macmillan.

——. 2007b. *Creating and Digitizing Language Corpora: Diachronic Databases Vol. 2*. Basingstoke: Palgrave Macmillan.

Bednarek, Monika. 2008. Semantic preference and semantic prosody re-examined. *Corpus Linguistics and Linguistic Theory* 4/2: 119–139.

Bell, Alan, Michelle L. Gregory, Jason M. Brenier, Daniel Jurafsky, Ayako Ikeno, and Cynthia Girand. 2002. Which predictability measures affect content word durations? *Proceedings of the Pronunciation Modeling and Lexicon Adaptation for Spoken Language Technology Workshop*, pp. 1–5.

Berg, Thomas. 2006. The internal structure of four-noun compounds in English and German. *Corpus Linguistics and Linguistic Theory* 2/2: 197–231.

Biber, Douglas. 1988. *Variation Across Speech and Writing*. Cambridge: Cambridge University Press.

——. 1990. Methodological issues regarding corpus-based analyses of linguistic variation. *Literary and Linguistic Computing* 5/4: 257–269.

——. 1993. Representativeness in corpus design. *Literary and Linguistic Computing* 8/4: 243–257.

Biber, Douglas, Susan Conrad, and Randi Reppen. 1998. *Corpus Linguistics: Investigating Language Structure and Use*. Cambridge: Cambridge University Press.

Bod, Rens, Jennifer Hay, and Stefanie Jannedy (eds.). 2003. *Probabilistic Linguistics*. Cambridge, MA: MIT Press.

Bortz, Jürgen. 2005. *Statistik für Human- und Sozialwissenschaftler.* 6th edn. Heidelberg: Springer.

Bowker, Lynne and Jennifer Pearson. 2002. *Working with Specialized Language: A Practical Guide to Using Corpora.* London and New York: Routledge.

Brinkmann, Ursula. 1997. *The Locative Alternation in German: Its Structure and Acquisition.* Amsterdam, Philadelphia: John Benjamins.

British National Corpus, version 2 (BNC World). 2001. Distributed by Oxford University Computing Services on behalf of the BNC Consortium. URL: <http://www.natcorp.ox.ac.uk/>

Brown, Roger. 1973. *A First Language: The Early Stages.* Cambridge, MA: Harvard University Press.

Bybee, Joan and Joanne Scheibman. 1999. The effect of usage on degrees of constituency: the reduction of *don't* in English. *Linguistics* 37/4: 575–596.

Church, Kenneth W. and Patrick Hanks. 1990. Word association norms, mutual information, and lexicography. *Computational Linguistics* 16/1: 22–29.

Church, Kenneth Ward, William Gale, Patrick Hanks, and Donald Hindle. 1991. Using statistics in lexical analysis. In *Lexical Acquisition: Exploiting On-line Resources to Build a Lexicon,* ed. Uri Zernik, 115–164. Hillsdale, NJ: Lawrence Erlbaum.

Church, Kenneth Ward, William Gale, Patrick Hanks, Donald Hindle, and Rosamund Moon. 1994. Lexical substitutability. In *Computational Approaches to the Lexicon,* ed. Beryl T. Sue Atkins and Antonio Zampolli, 153–177. Oxford: Oxford University Press.

Clark, Herbert H. and Jean E. Fox Tree. 2002. Using *uh* and *um* in spontaneous speaking. *Cognition* 84/1: 73–111.

Conrad, Susan. 2000. Will corpus linguistics revolutionize grammar teaching in the 21st century? *TESOL Quarterly* 34: 548–560.

Crawley, Michael J. 2002. *Statistical Computings: An Introduction to Data Analysis Using S-Plus.* Chichester: John Wiley and Sons.

——. 2005. *Statistics: An Introduction using R.* Chichester: John Wiley and Sons.

——. 2007. *The R Book.* Chichester: John Wiley and Sons.

Dalbey, John. 2003. Pseudocode standard. Downloaded from <http://www.csc.calpoly.edu/~jdalbey/SWE/pdl_std.html> on November 26, 2006.

Dalgaard, Peter. 2002. *Introductory Statistics with R.* Berlin, Heidelberg: Springer.

Danielsson, Pernilla. 2004. Simple Perl programming for corpus work. In *How to Use Corpora in Language Teaching,* ed. John McHardy Sinclair, 225–246. Amsterdam and Philadelphia: John Benjamins.

Diessel, Holger. 1996. Processing factors of pre- and postposed adverbial clauses. *Proceedings of the 22nd Annual Meeting of the Berkeley Linguistics Society,* pp. 71–82.

——. 2001. The ordering distribution of main and adverbial clauses: a typological study. *Language* 77/3: 343–365.

——. 2005. Competing motivations for the ordering of main and adverbial clauses. *Linguistics* 43/3: 449–470.

Diessel, Holger and Michael Tomasello. 2005. Particle placement in early child language: a multifactorial analysis. *Corpus Linguistics and Linguistic Theory* 1/1: 89–112.

Dilts, Philip and John Newman. 2006. A note on quantifying "good" and "bad" prosodies. *Corpus Linguistics and Linguistic Theory* 2/2: 233–242.

Divjak, Dagmar and Stefan Th. Gries. 2006. Ways of trying in Russian: clustering behavioral profiles. *Corpus Linguistics and Linguistic Theory* 2/1: 23–60.

Ellis, Nick C. 2002a. Frequency effects in language processing and acquisition. *Studies in Second Language Acquisition* 24/2: 143–188.

——. 2002b. Reflections on frequency effects in language processing. *Studies in Second Language Acquisition* 24/2: 297–339.

Estival, Dominique. 1985. Syntactic priming of the passive in English. *Text* 5/1: 7–22.

Evert, Stefan. 2004. The statistics of word cooccurrences: word pairs and collocations. Ph.D. diss., University of Stuttgart.

Forta, Ben. 2004. *Sams Teach Yourself Regular Expressions in 10 Minutes.* Indianapolis, IN: Sams Publishing.

Fox, Gwyneth. 1998. Using corpus data in the classroom. In *Materials Development in Language Teaching,* ed. Brian Tomlinson, 25–43. Cambridge: Cambridge University Press.

Fox Tree, Jean E. and Herbert H. Clark. 1997. Pronouncing *the* as *thee* to signal problems in speaking. *Cognition* 62/2: 151–167.

Friedl, Jeffrey E.F. 2006. *Mastering Regular Expressions*. 3rd edn. Cambridge, MA: O'Reilly.

Gahl, Susanne and Susan Marie Garnsey. 2004. Knowledge of grammar, knowledge of usage: syntactic probabilities affect pronunciation variation. *Language* 80/4: 748–775.

Givón, Talmy. 1992. The grammar of referential coherence as mental processing instructions. *Linguistics* 30/1: 5–55.

Good, Nathan A. 2005. *Regular Expression Recipes: A Problem-solution Approach*. Berkeley, CA: Apress.

Good, Philip I. 2005. *Introduction to Statistics through Resampling Methods and R/S-Plus*. Hoboken, NJ: John Wiley and Sons.

Gries, Stefan Th. 2001. A corpus-linguistic analysis of -*ic* and -*ical* adjectives. *ICAME Journal* 25: 65–108.

——. 2002a. Evidence in linguistics: three approaches to genitives in English. In *LACUS Forum XXVIII: what constitutes evidence in linguistics?*, ed. Ruth M. Brend, William J. Sullivan, and Arle R. Lommel, 17–31. Fullerton, CA: LACUS.

——. 2002b. Preposition stranding in English: predicting speakers' behaviour. In *Proceedings of the Western Conference on Linguistics. Vol. 12*, ed. Vida Samiian, 230–241. California State University, Fresno, CA.

——. 2003a. *Multifactorial Analysis in Corpus Linguistics: a study of particle placement*. London and New York: Continuum Press.

——. 2003b. Testing the sub-test: a collocational-overlap analysis of English -*ic* and -*ical* adjectives. *International Journal of Corpus Linguistics* 8/1: 31–61.

——. 2004a. *Coll.analysis 3*. A program for R for Windows 2.x.

——. 2004b. Isn't that fantabulous? How similarity motivates intentional morphological blends in English. In *Language, Culture, and Mind*, ed. Michel Achard and Suzanne Kemmer, 415–428. Stanford, CA: CSLI.

——. 2005a. Syntactic priming: a corpus-based approach. *Journal of Psycholinguistic Research* 34/4: 365–399.

——. 2005b. Null-hypothesis significance testing of word frequencies: a follow-up on Kilgarriff. *Corpus Linguistics and Linguistic Theory* 1/2: 277–294.

——. 2006a. Some proposals towards more rigorous corpus linguistics. *Zeitschrift für Anglistik und Amerikanistik* 54/2: 191–202.

——. 2006b. Cognitive determinants of subtractive word-formation processes: a corpus-based perspective. *Cognitive Linguistics* 17/4: 535–558.

——. 2007a. Exploring variability within and between corpora: some methodological considerations. *Corpora* 1/2: 109–151.

——. 2007b. Towards a more useful measure of dispersion in corpus data. Paper presented at Corpus Linguistics.

——. 2008a. Corpus-based methods in analyses of SLA data. In *Handbook of Cognitive Linguistics and Second Language Acquisition*, ed. Peter Robinson and Nick Ellis, 406–431. New York: Routledge.

——. 2008b. Phraseology and linguistic theory: a brief survey. In *Phraseology: An Interdisciplinary Perspective*, ed. Sylviane Granger and Fanny Meunier, 3–25. Amsterdam and Philadelphia: John Benjamins.

——. 2008c. Dispersions and adjusted frequencies in corpora. *International Journal of Corpus Linguistics*.

——. to appear. *Statistics for Linguistics Using R: A Practical Introduction*. Berlin, New York: Mouton de Gruyter.

Gries, Stefan Th. and Caroline V. David. 2007. This is kind of/sort of interesting: variation in hedging in English. In *Proceedings of ICAME 27*, ed. Päivi Pahta, Irma Taavitsainen, and Terttu Nevalainen. eVARIENG: Methodological Interfaces, University of Helsinki.

Gries, Stefan Th. and Dagmar Divjak. to appear. Behavioral profiles: a corpus-based approach towards cognitive semantic analysis. In *New Directions in Cognitive Linguistics*, ed. Vyvyan Evans and Stephanie Pourcel. Amsterdam and Philadelphia: John Benjamins.

Gries, Stefan Th., Beate Hampe, and Doris Schönefeld. 2005. Converging evidence: bringing together experimental and corpus data on the association of verbs and constructions. *Cognitive Linguistics* 16/4: 635–676.

——. to appear. Converging evidence II: more on the association of verbs and constructions. In *Empirical and*

Experimental Methods in Cognitive/Functional Research, ed. John Newman and Sally Rice. Stanford, CA: CSLI.

Gries, Stefan Th. and Anatol Stefanowitsch. 2004a. Extending collostructional analysis: a corpus-based perspective on "alternations". *International Journal of Corpus Linguistics* 9/1: 97–129.

——. 2004b. Co-varying collexemes in the *into*-causative. In *Language, Culture, and Mind*, ed. Michel Achard and Suzanne Kemmer, 225–236. Stanford, CA: CSLI.

—— (eds.). 2006. *Corpora in Cognitive Linguistics: Corpus-Based Approaches to Syntax and Lexis*. Berlin, New York: Mouton de Gruyter.

Gries, Stefan Th. and Sabine Stoll. to appear. Finding developmental groups in acquisition data: variability-based neighbor clustering, *Journal of Quantitative Linguistics* 16/3.

de Haan, Pieter. 1992. The optimum corpus sample size. In *New Directions in English Language Corpora*, ed. Gerhard Leitner, 3–19. Berlin and New York: Mouton de Gruyter.

Hearst, Marti A. 1997. TextTiling: Segmenting text into multi-paragraph subtopic passages. *Computational Linguistics* 23/1: 33–64.

Heatley, Alex, Paul I.S. Nation, and Avery Coxhead. 2002. RANGE and FREQUENCY programs. <http://www.vuw.ac.nz/lals/staff/Paul_Nation>

Höche, Silke. 2009. *Cognate object constructions in English: a cognitive linguistic account*. Tübingen: Gunter Narr.

Hoffmann, Thomas. 2006. Corpora and introspection as corroborating evidence: the case of preposition placement in English relative clauses. *Corpus Linguistics and Linguistic Theory* 2/2: 165–195.

Hofland, Knud and Stig Johansson. 1982. *Word Frequencies in British and American English*. Bergen: Norwegian Computing Centre for the Humanities/London: Longman.

Hundt, Marianne, Nadja Nesselhauf, and Carolin Biewer (eds.). 2007. *Corpus Linguistics and the Web*. Amsterdam, New York, NY: Rodopi.

Hunston, Susan. 2002. *Corpora in Applied Linguistics*. Cambridge: Cambridge University Press.

Hunston, Susan and Gill Francis. 2000. *Pattern Grammar: A Corpus-driven Approach to the Lexical Grammar of English*. Amsterdam and Philadelphia: John Benjamins.

Jackendoff, Ray. 1997. Twistin' the night away. *Language* 73/3: 534–559.

Johansson, Stig. 1980. The LOB corpus of British English texts: presentation and comments. *ACCL Journal* 1: 25–36.

Johansson, Stig and Knut Hofland. 1989. *Frequency Analysis of English Vocabulary and Grammar, Based on the LOB Corpus*. Oxford: Clarendon Press.

Johnson, Keith. 2008. *Quantitative Methods in Linguistics*. Malden, MA and Oxford: Blackwell.

Jones, Steven. 2002. *Antonymy: A Corpus-based Perspective*. London and New York: Routledge.

Joseph, Brian D. 2004. On change in Language and change in language. *Language* 80/3: 381–383.

Juola, Patrick. 2008. *Authorship Attribution*. Hanover, MA and Delft: Now Publishers.

Jurafsky, Daniel, Alan Bell, Michelle Gregory and William D. Raymond. 2000. Probabilistic relations between words: Evidence from reduction in lexical production. In *Frequency and the Emergence of Linguistic Structure*, ed. Joan Bybee and Paul Hopper, 229–254. Amsterdam and Philadelphia: John Benjamins.

Jurafsky, Daniel and James H. Martin. 2000. *Speech and Language Processing*: [. . .]. Upper Saddle River, NJ: Prentice Hall.

Justeson, John S. and Slava M. Katz. 1991. Co-occurrence of antonymous adjectives and their contexts. *Computational Linguistics* 17/1: 1–19.

——. 1995. Principled disambiguation: discriminating adjective senses with modified nouns. *Computational Linguistics* 21/1: 1–27.

Kaunisto, Mark. 1999. *Electric/electrical* and *classic/classical*: variation between the suffixes -*ic* and -*ical*. *English Studies* 80/4: 343–370.

——. 2001. Nobility in the history of adjectives ending in -*ic* and -*ical*. In *LACUS Forum XXVII: speaking and comprehending*, ed. Ruth Brend, Alan K. Melby, and Arle R. Lommel, 35–46. Fullerton, CA: LACUS.

Kay, Paul and Charles J. Fillmore. 1999. Grammatical constructions and linguistic generalizations: the What's X doing Y? construction. *Language* 75/1: 1–33.

Keller, Frank. 2000. Gradience in grammar: experimental and computational aspects of degrees of grammaticality. Ph. D. diss., University of Edinburgh.

Kennedy, Graeme. 1998. *An Introduction to Corpus Linguistics*. London and New York: Longman.

Kilgarriff, Adam. 2001. Comparing corpora. *International Journal of Corpus Linguistics* 6/1: 1–37.

——. 2005. Language is never, ever, ever random. *Corpus Linguistics and Linguistic Theory* 1/2: 263–276.

Kilgarriff, Adam and David Tugwell. 2001. Word Sketch: extraction and display of significant collocations for lexicography. *Proceedings of the Workshop on Collocations*, ACL, pp. 32–38.

Kolari, Pranam and Anupam Joshi. 2004. Web mining: research and practice. *IEEE Computing in Science and Engineering* 6/4: 49–53.

Laufer, Batia. 2005. Lexical Frequency Profiles: from Monte Carlo to the real world. A response to Meara. *Applied Linguistics* 26/4: 581–587.

Laufer, Batia and Paul Nation. 1995. Vocabulary size and use: lexical richness in L2 written production. *Applied Linguistics* 16/3: 307–322.

Leech, Geoffrey N. 1992. Corpora and theories of linguistic performance. In *Directions in Corpus Linguistics*, ed. Jan Svartvik, 105–122. Berlin and New York: Mouton de Gruyter.

——. 1993. Corpus annotation schemes. *Literary and Linguistic Computing* 8/4: 275–281.

Leech, Geoffrey and Roger Fallon. 1992. Computer corpora: what do they tell us about culture? *ICAME Journal* 16: 29–50.

Levin, Beth. 1993. *Verb Classes and Alternations: A Preliminary Investigation*. Chicago, IL: University of Chicago Press.

Lewis, Michael (ed.). 2000. *Teaching Collocation: Further Developments in the Lexical Approach*. Hove: Language Teaching Publications.

Ligges, Uwe. 2005. *Programmieren mit R*. Berlin and Heidelberg: Springer.

Maindonald, John and John Braun. 2003. *Data Analysis and Graphics Using R: An Example-based Approach*. Cambridge: Cambridge University Press.

Manning, Christopher D. and Hinrich Schütze. 2000. *Foundations of Statistical Natural Language Processing*. 2nd printing with corrections. Cambridge, MA and London: MIT Press.

Mason, Oliver. 1997. The weight of words: an investigation of lexical gravity. *Proceedings of PALC*, pp. 361–375.

——. 1999. Parameters of collocation: the word in the centre of gravity. In *Corpora Galore: Analyses and Techniques in Describing English*, ed. John Kirk, 267–280. Amsterdam: Rodopi.

McEnery, Tony and Andrew Wilson. 2003. *Corpus Linguistics*. 2nd edn. Edinburgh: Edinburgh University Press.

McEnery, Tony, Richard Xiao, and Yukio Tono. 2006. *Corpus-based Language Studies: An Advanced Resource book*. London and New York: Routledge.

Meara, Paul. 2005. Lexical frequency profiles: a Monte Carlo analysis. *Applied Linguistics* 26/1: 32–47.

Meyer, Charles F. 2002. *English Corpus Linguistics: An Introduction*. Cambridge: Cambridge University Press.

Mukherjee, Joybrato. 2007. Corpus linguistics and linguistic theory: general nouns and general issues. *International Journal of Corpus Linguistics* 12/1: 131–147.

Murrell, Paul. 2005. *R Graphics*. Boca Raton, FL: Chapman and Hall.

Oakes, Michael P. 1998. *Statistics for Corpus Linguistics*. Edinburgh: Edinburgh University Press.

——. 2003. Text categorisation: automatic discrimination between US and UK English using the chi-square test and high ratio pairs. Downloaded from <http://www.cet.sunderland.ac.uk/IR/oakesRL2003.pdf> on April 5, 2006.

Oh, Sun-Young. 2000. *Actually* and in *fact* in American English: a data-based analysis. *English Language and Linguistics* 4/2: 243–268.

Okada, Sadayuki. 1999. On the function and distribution of the modifiers *respective* and *respectively*. *Linguistics* 37/5: 871–903.

O'Keeffe, Anna, Michael McCarthy, and Ronald Carter. 2007. *From Corpus to Classroom*. Cambridge: Cambridge University Press.

Panther, Klaus-Uwe and Linda L. Thornburg. 2002. The roles of metaphor and metonymy in English *-er* nominals. In *Metaphor and Metonymy in Comparison and Contrast*, ed. René Dirven and Ralf Pörings, 279–319. Berlin and New York: Mouton de Gruyter.

Partington, Alan. 1998. *Patterns and Meanings: Using Corpora for English Language Research and Teaching.* Amsterdam and Philadelphia: John Benjamins.

Pullum, Geoffrey K. 2006. Ungrammaticality, rarity, and corpus use. *Corpus Linguistics and Linguistic Theory* 3/1: 33–47.

Quirk, Randolph, Sidney Greenbaum, Geoffrey, and Jan Svartvik. 1985. *A Comprehensive Grammar of the English Language.* London, New York: Longman.

R Development Core Team. 2008. *R: A language and environment for statistical computing.* R Foundation for Statistical Computing, Vienna, Austria. ISBN 3-900051-07-0, URL <http://www.R-project.org>

Rayson, Paul E. and Roger Garside. 2000. Comparing corpora using frequency profiling. *Proceedings of the Workshop on Comparing Corpora*, pp. 1–6.

Rayson, Paul E., Geoffrey Leech, and Mary Hodges. 1997. Social differentiation in the use of English vocabulary: some analysis of the conversational component of the British National Corpus. *International Journal of Corpus Linguistics* 2/1: 133–152.

Rayson, Paul, Andrew Wilson, and Geoffrey N. Leech. 2001. Grammatical word class variation within the British National Corpus Sampler. In *New Frontiers of Corpus Research*, ed. Pam Peters, Peter Collins, and Adam Smith, 295–306. Amsterdam: Rodopi.

Roland, Douglas and Daniel Jurafsky. 1998. How verb subcategorization frequencies are affected by corpus choice. *Proceedings of the 17th International Conference on Computational Linguistics*—Volume 2, pp. 1,122–1,128.

Ryder, Mary Ellen. 1999. *Bankers* and *blue-chippers*: an account of -er formations in present-day English. *English Language and Linguistics* 3/2: 269–297.

Sankoff, David and Suzanne Laberge. 1978. Statistical dependence among successive occurrences of a variable in discourse. In *Linguistic Variation: Models and Methods*, ed. David Sankoff, 119–126. New York: Academic Press.

Schmid, Hans-Jörg. 2000. *English Abstract Nouns as Conceptual Shells: From Corpus to Cognition.* Berlin and New York: Mouton de Gruyter.

Schönefeld, Doris. 1999. Corpus linguistics and cognitivism. *International Journal of Corpus Linguistics* 4/1:131–171.

Sharoff, Serge. 2006. Open-source corpora: using the net to fish for linguistic data. *International Journal of Corpus Linguistics* 11/4: 435–462.

Sinclair, John McHardy (ed.). 1987. *Looking Up: An Account of the COBUILD Project in Lexical Computing.* London: Collins.

Sinclair, John McHardy. 1991. *Corpus, Concordance, Collocation.* Oxford: Oxford University Press.

——. 2003. *Reading Concordances: An Introduction.* London: Pearson.

——. 2004. *Trust the Text: language, corpus, and discourse.* London, New York: Routledge.

Spector, Phil. 2008. *Data Manipulation with R.* New York: Springer.

Stefanowitsch, Anatol. 2005. New York, Dayton (Ohio), and the raw frequency fallacy. *Corpus Linguistics and Linguistic Theory* 1/2: 295–301.

——. Negative evidence and the raw frequency fallacy. *Corpus Linguistics and Linguistic Theory* 2/1: 61–77.

Stefanowitsch, Anatol and Stefan Th. Gries. 2003. Collostructions: investigating the interaction between words and constructions. *International Journal of Corpus Linguistics* 8/2: 209–243.

——. 2005. Covarying collexemes. *Corpus Linguistics and Linguistic Theory* 1/1: 1–43.

Stefanowitsch, Anatol and Stefan Th. Gries (eds.). 2006. *Corpus-based Approaches to Metaphor and Metonymy.* Berlin and New York: Mouton de Gruyter.

Steinberg, Danny D. 1993. *An Introduction to Psycholinguistics.* London and New York: Longman.

Stevens, Vance. 1991. Classroom concordancing: vocabulary materials derived from relevant, authentic text. *English for Specific Purposes* 10/1: 35–46.

Stubblebine, Tony. 2003. *Regular Expression Pocket Reference.* Cambridge, MA: O'Reilly.

Stubbs, Michael. 1995. Collocations and semantic profiles: on the cause of the trouble with quantitative studies. *Functions of Language* 2/1: 23–55.

——. 2001. *Words and Phrases: Corpus Studies in Lexical Semantics.* Oxford: Blackwell.

Szmrecsanyi, Benedikt. 2005. Language users as creatures of habit: a corpus-based analysis of persistence in spoken English. *Corpus Linguistics and Linguistic Theory* 1/1: 113–149.

Tao, Hongyin. 2007. A corpus-based investigation of *absolutely* and related phenomena in spoken American English. *Journal of English Linguistics* 35/1: 5–29.

Tao, Hongyin and Charles F. Meyer. 2006. Gapped coordinations in English: form, usage, and implications for linguistic theory. *Corpus Linguistics and Linguistic Theory* 2/2: 129–163.

Tummers, José, Kris Heylen, and Dirk Geeraerts. 2005. Usage-based approaches in cognitive linguistics: a technical state of the art. *Corpus Linguistics and Linguistic Theory* 1/2: 225–261.

von Eye, Alexander. 1990. *Introduction to Configural Frequency Analysis: The Search for Types and Antitypes in Cross-classifications.* Cambridge: Cambridge University Press.

Watt, Andrew. 2005. *Beginning Regular Expressions.* Indianapolis, IN: Wiley Publishing.

Whitsitt, Sam. 2005. A critique of the concept of semantic prosody. *International Journal of Corpus Linguistics* 10/3: 283–305.

Wiechmann, Daniel. 2008. On the computation of collostruction strength: testing measures of association as expressions of lexical bias. *Corpus Linguistics and Linguistic Theory* 4/2: 253–290.

Wiechmann, Daniel and Stefan Fuhs. 2006. Concordancing software. *Corpus Linguistics and Linguistic Theory* 2/1: 109–127.

Wulff, Stefanie. 2003. A multifactorial corpus analysis of adjective order in English. *International Journal of Corpus Linguistics* 8/2: 245–282.

Wynne, Martin (ed.). 2005. *Developing Linguistic Corpora: a Guide to Good Practice.* Oxford: Oxbow Books, <http://ahds.ac.uk/linguistic-corpora/>

Xiao, Richard and Tony McEnery. 2006. Collocation, semantic prosody, and near synonymy: a cross-linguistic perspective. *Applied Linguistics* 27/1: 103–129.

Youmans, Gilbert. 1990. Measuring lexical style and competence: the type–token vocabulary curve. *Style* 24/4: 584–599.

——. 1991. A new tool for discourse analysis: the Vocabulary Management Profile. *Language* 67/4: 763–789.

Endnotes

1. Introduction

1 It is worth pointing out, though, that many corpus-linguistic findings and the more theoretical interpretation of what these findings reveal about human linguistic knowledge and processing are so compatible with positions held in contemporary usage-based linguistics, cognitive linguistics, and construction grammar that these may be considered as the most useful theoretical background against which to interpret corpus-linguistic findings of all sorts. In fact, the compatibility is so large that it is in fact surprising that there is so little cooperation between the two fields; cf. Leech (1992), Schönefeld (1999), Schmid (2000), or Gries (2008b) for more detailed discussion.

2. The Three Central Corpus-linguistic Methods

1 It is only fair to mention, however, that (i) error collections have proven extremely useful in spite of what, from a strict corpus-linguistic perspective, may be considered shortcomings, and that (ii) compilers of corpora of lesser-spoken languages such as typologists investigating languages with few written records suffer from just the same data scarcity problems.

2 One important characteristic of language is reflected in nearly all frequency lists. When you count how often each word form occurs in a particular corpus, you usually get a skewed distribution such that:

(i) a few word forms—usually short function words—account for the lion's share of all word forms (for example, in the Brown corpus, the ten most frequent word forms out of all approximately 41,000 different word forms already account for nearly 24 percent of all word forms); and

(ii) most word forms occur rather infrequently (for example, 16,000 of the word forms in the Brown corpus occur only once).

These observations are a subset of Zipf's laws, the most famous of which states that the frequency of any word is approximately inversely proportional to its rank in a frequency list.

3 Even if the term *stop list* is new to you, you are probably already familiar with the concept: if you enter the following search terms into Google *host of problems* (without quotes), Google tells you that you get results for *host problems* only.

4 Note, however, that this may turn into an interesting finding once you find that some adjectives exhibit a tendency for such kinds of additional premodification while others don't. Also, note in passing that in the context of regular expressions (cf. Section 3.7.5 and following), *precision* and *recall* are sometimes referred to as *specificity* and *sensitivity* respectively.

5 Strictly speaking, a concordance does not have to list every occurrence of a word in a corpus. Some programs output only a user-specified number of occurrences, usually either the first *n* occurrences in the corpus or *n* randomly chosen occurrences from the corpus. Also, a rarer use of the word *concordance* does not refer to a list of (all) occurrences of a particular word together with a user-specified context, but to such a list of all word forms.

3. An Introduction to R

1 One important point needs to be mentioned here. For reasons having to do with the difficulty of getting copyright clearances from publishers, most files to be used here are either publicly available (e.g. Wikipedia entries, the GNU Public License, chat files from the CHILDES database, etc.) or are randomly assembled and manufactured from authentic corpus files. For example, the files called <bnc*.txt> are not real files from the BNC but were assembled by putting together randomly chosen and post-processed individual lines from said corpus; the same is true of the data from the Brown corpus, and the ICE-GB. While this takes care of the copyright issues, two bad consequences are that we cannot look at discourse-phenomena that require longer stretches of cohesive discourse and that, in one case—the SGML/XML versions of the BNC—the annotation is not completely identical with that of the original corpus. However, with regard to the former, I hope you will agree later that this book still offers a rather diverse set of topics. With regard to the latter, I have made sure that the formatting is similar enough for you to be able to handle the real corpus files if you or your department obtains a copy.

2 Enter `iconvlist()`¶ at the R prompt for a list of encodings supported on your computer.

3 We will talk about the meaning of "[1]" in Section 3.2 below.

4 Of course, you will most likely get different values since we are generating *random* numbers.

5 If you let R output vectors to the screen that unlike the above are so long that they cannot be shown in one row anymore, R will use as many lines as necessary to represent all elements of the vector, and each line will start with the number (in square brackets) of the first element of the vector that is given in each line. Try it out: enter `1:100`¶ at the prompt.

6 Since Microsoft Windows is still, for better or for worse, the most widely used operating system, below and in the scripts, I will mostly use the function `choose.files` (in the scripts with the argument `default="..."` providing the path of the relevant file). On Windows, this is more convenient than `file .choose()` and allows you to just press ENTER to access the data if you stored the files in the recommended locations.

7 As a Perl user, you must bear in mind that while Perl begins to number elements of an array at 0, R does not—it begins at 1.

8 This is actually a generally important concept. R shows missing or not available data as NA. Since this may happen for many different reasons, let me mention two aspects relevant in the present context. First, you can check whether something is not available using the function `is.na`, which takes the element to be tested as its argument. For example, if the vector a consists only of the number 2, then `is.na(a)`¶ gives FALSE (because there is something, the number 2)¶, but `is.na(a[2])`¶ gives TRUE, because there is no second element in a. Second, many functions have default settings for how to deal with NA, which is why the unexpected behavior of a function can often be explained by checking whether NAs distort the results. To determine the default behavior of functions regarding NA, check their documentation.

9 Note in passing that you can only change *vectors* by inserting new values. With *factors*, you must first define the new level and then you can assign it to any element:

```
>·levels(Class)<-append(levels(Class),·"unclear")¶
>·Class[4]<-"unclear"¶
```

10 Microsoft Excel 2007 now allows users to load files with more than one million rows, for OpenOffice.org 3.0 Calc, this capability has been added to a wish list.

11 Note that R is vastly superior to some common spreadsheet software and also most concordancing software, since the number of columns according to which you can sort is in principle unlimited.

12 It is worth pointing out that, even though the following code is a very clumsy way of obtaining the desired sums, it still works! The beauty of R is that while there are many very convenient functions to achieve things and ingenious solutions to tricky problems, you need not attain great proficiency before you can write your own code. Thus, don't hold back in trying things out.

13 Many concordancers also offer regular expressions, but surprisingly not all of them do: some, including the widely distributed WordSmith Tools 4, offer only a limited set of wildcards (cf. Wiechmann and Fuhs 2006).

14 The issue of how paragraph breaks are represented in files is actually more problematic than it might appear at first sight. Hitting ENTER in a text document generates a paragraph break so that the user continues to enter data on a new line. Some programs' display, such as Microsoft Word or OpenOffice.org

Writer, can be configured to show formatting marks and usually display the Pilcrow sign "¶" as a paragraph mark. However, on computers running Microsoft Windows OS what is actually stored are two characters: a carriage return (CR) and a line feed (LF). In many programming languages, text editors, and also in R, the former is represented as "\r", while the latter—often also referred to as newline—is represented as "\n". Unfortunately, other operating systems use different characters to represent what for the user is a paragraph break: Unix systems use only LF and older Macintosh systems only use CR. In order to avoid errors with regular expressions, it is often useful to look at the file in a text editor that displays all control characters to determine which paragraph mark is actually used, as in Figure 3.6 shown below, where the upper and the lower panel show a Windows and a Unix file respectively.

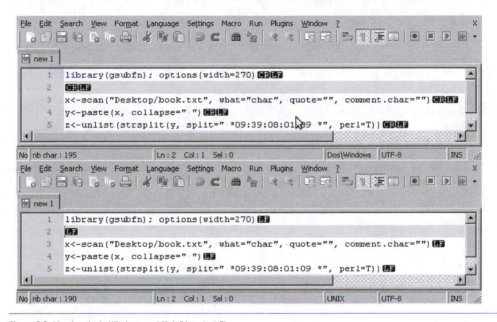

Figure 3.6 Line breaks in Windows and Unix/Linux text files.

15 In a language with a many-to-many correspondence of spelling and pronunciation such as English, this can of course only be a heuristic. However, this regular expression would certainly narrow down the search space considerably, and when applied to more regular languages such as Spanish or to phonetic transcription, which could in fact be assigned to a text given a database such as CELEX, the regular expression below could do the trick. An alternative area of application would be to use this, for example, to retrieve pairs of adverbs in English (because most adverbs end in -ly). Note also that this is again an example of something that regular concordancing software usually cannot do.

16 This function can in fact take arguments, but we will not deal with them here because they mostly have to do with customizing the file choice and the selection dialog; enter ?choose.files¶ at the R prompt for details.

17 This by the way is the second alternative for choose.files() for Macintosh and Linux users, but of course it presupposes that the relevant files have a consistent pattern.

18 This function can therefore be used to approximate some of the functionality of the File Utilities in WordSmith Tools 5 (such as comparing files).

4. Using R in Corpus Linguistics

1 This cleaning process could also be done after the loop. We included it in the loop in order to minimize the amount of elements to be stored in the whole.corpus.

2 This is the format of the corpus on the ICAME CD-ROM. In the freely available NLTK-version (cf. the Appendix for the link) a slash is used instead of an underscore.

3 This is more relevant than it may seem. For example, Berg (2006) retrieves British English and German four-noun compounds and explicitly comments on how such cases had to be discarded manually: "The 1152 items were reduced by eliminating all cases where [...] a sentence or paragraph boundary occurred

within the 'compound' (e.g. the first two nouns formed the end of the newspaper headline and the last two the beginning of the running text)" (Berg 2006: 208).

4 Two other rather ingenious alternatives were suggested to me by Gabor Grothendieck (p.c.). They apply to a vector v with one character string into it to explore the documentation for embed and head/tail:

```
>•alternative.1<-tail(embed(strsplit(v,•"•")[[1]],•2),•-1)[,2:1]¶
>•alternative.2<-strsplit(v,•"•")[[1]][-1];•cbind(head(alternative.2,•-1),•
   tail(alternative.2,•-1))¶
```

5 Corpora with a somewhat similar format—in the sense of having one word per line—are Christine and Susanne or versions of the Switchboard corpus.

6 It is worth pointing out that while the correct saving of the output is an elementary function, commercial concordancing software can fail to do this . . . As Wiechmann and Fuhs (2006:123, note 1) show, Mono-Conc Pro 2.2, for example, cannot properly save very long matches.

7 Strictly speaking, this statement is incorrect; we will correct it presently.

8 The durations were measured with the function system.time, using 100 replications.

9 Yet other corpora have standalone annotation; e.g. the American National Corpus, but I will not discuss that here.

10 The hexadecimal system is a numerical system that has 16 as its base (unlike the familiar decimal system, which of course has base 10). Instead of the 10 numbers 0 to 9 that the decimal system uses, the hexadecimal system uses 16 distinct characters, 0–9 and a–f (or A–F). Thus, the decimal number 8 is also 8 in the hexadecimal system, the decimal number 15 is hexadecimal F, the decimal number 23 is hexadecimal 17 (namely $1*16 + 7*1$), and the decimal number 1970 is hexadecimal 7B2 (namely $7*256 + 11*16 + 2*1$).

5. Some Statistics for Corpus Linguistics

1 I will not deal with other kinds of variables such as confounding or moderator variables.

2 Enter this into R and look at the help for dbinom: sum(dbinom(60:100,·100,·0.5))¶.

3 Note that this result is not proving that I was cheating in the mathematical or logical sense of the word *proof*, because there is still a small chance—the 0.02844397 we computed—that the result is due to chance. The obtained result is just not very likely if I have not cheated.

 Note in passing that there is a quasi standard way of reporting results: when you adopt a significance level of 0.05, then *p*-values smaller than 0.05, but larger than or equal to 0.01 are referred to as "significant", *p*-values smaller than 0.01, but larger than or equal to 0.001 are called "very significant", and *p*-values smaller than 0.001 are referred to as "highly significant".

4 Recall from Section 3.8 above how to save graphs.

5 As a matter of fact, 5 is the most widely cited threshold value. There is, however, a variety of studies that indicate that the chi-square test is fairly robust even if some expected frequencies are smaller than 5, especially when the null hypothesis is a uniform distribution (as in the present case); cf. e.g. Zar (1999: 470) for a short summary. In order not to complicate matters here any further, I stick to the conservative threshold and advise the reader to consult statistics textbooks for more comprehensive coverage.

 If whatever threshold value adopted on the basis of the pertinent literature is not met, you should use either the binomial test (if you have just two categories, as in the present example; cf. ?binom.test¶ or ?dbinom¶) or the multinomial test (for 3+ categories; cf. ?dmultinom¶).

6 For further options, enter ?chisq.test¶ into R.

7 If whatever threshold value adopted on the basis of the pertinent literature is not met and if you have 3+ categories, you might want to consider merging variable levels so that the expected frequencies are increased. If this does not make sense or if you only have two categories to begin with, a bootstrapping procedure may be a useful alternative; cf. the help for chisq.test in R, in particular the argument simulate.p.value.

8 Strictly speaking, the theoretical range from 0 to 1 can only be obtained for particular distributions, but it is nevertheless a good heuristic for the interpretation of the result. Also, there is no uniformly accepted rule when to call an effect weak, moderate, or strong: different authors have suggested different scales.

9 If one's data set involves just such a table—only two variables—one might create such a table manually. To that end, use the function matrix with three arguments: (i) the vector containing all frequencies to be

entered into the matrix, (ii) byrow=TRUE or byrow=FALSE, depending on whether you enter the numbers in the vector row-wise or column-wise; and (iii) ncol=… or nrow=… to provide the number of columns or rows. Thus:

```
>•Gries.2003<-matrix(c(125,•64,•69,•145),•byrow=TRUE,•ncol=2)¶
>•chisq.test(Gries.2003,•correct=F)¶
```

This is a handy way of creating tables for subsequent chi-square tests if you cannot or do not want to read in new data files, and you will need to use this method again below. However, this only arranges the frequencies accordingly, but it does not provide row and column names, which we need for a nice (publishable) plot. Thus, you should also add row and column names.

```
>•rownames(Gries.2003)<-c("V_Part_DO",•"V_DO_Part")¶
>•colnames(Gries.2003)<-c("abstract",•"concrete")¶
```

Since this is then already a matrix, you can apply assocplot directly (but you need to give the labels for the rows and columns in the plot because R does not know them—recall that you did not create this table by having R infer these labels from the column headers):

```
>•assocplot(Gries.2003,•xlab="CONSTRUCTION",•ylab="CONCRETENESS")¶
```

10 A program to compute many association measures from tabular data, Coll.analysis 3, is available from me upon request.

11 A program to compute hierarchical configural frequency analyses, HCFA 3.2, is available from me upon request.

12 The same is actually true of corpus frequencies. Too many studies contend themselves by just reporting frequencies of occurrence or co-occurrence only without also reporting a measure of corpus dispersion, which can sometimes hide a considerable bias in the data; cf. Gries (2006a, to appear b) for discussion.

13 Strictly speaking, the default version of the *t*-test for independent samples that is implemented in R does not require homogeneity of variances, but there is still the violation of normality.

14 If your data are from English, German, or Dutch, this book may have actually taught you enough for you to be able to automatize the task of counting syllables of words using the information provided in the CELEX database (cf. Baayen, Piepenbrock, and Gulikers 1995). You can read in files from the CELEX database files with scan, use strsplit to turn the file into a list and then use sapply to get at individual fields.

15 As a matter of fact, all of these approaches (including logistic regression) are special cases of general(ized) linear modeling.

16 If you have more than one case study, then each case study gets its own methods, results, and discussion sequence, and then you pull together the results of all case studies in one section entitled "General discussion" at the end of the paper.

Index